BASIC TRANSISTOR COURSE

2ND EDITION

BY STAN GIBILISCO

TAB BOOKS Inc.
BLUE RIDGE SUMMIT, PA. 17214

SECOND EDITION

FIRST PRINTING

Printed in the United States of America

Library of Congress Cataloging in Publication Data

Gibilisco, Stan.
Basic transistor course.

Rev. ed. of: Basic transistor course / Paul Rodger Kenian. (1962)
Includes index.
1. Transistors. 2. Transistor circuits.
I. Kenian, Paul Rodger. Basic transistor course.
II. Title.
TK7871.9.G45 1984 621.3815'28 83-27193
ISBN 0-8306-0105-8
ISBN 0-8306-0605-X (pbk.)

Contents

Introduction

COLUMBUS IS POPULARLY KNOWN AS A HARDY AND VENturesome explorer who tested his beliefs in a very practical manner, but long before Christopher, other lesser known adventurers probed the American coastline. And, in the same way, while John Bardeen and Walter H. Brattain of Bell Telephone Laboratories are deservedly credited with the invention of the transistor, many others, some long before these two, poked, probed, and tested the very curious and intriguing properties of crystalline materials.

As far back as 1855, and that is really reaching into the far recesses of the beginnings of electronics, some of the basic properties of semiconductors had been investigated. Even before then in 1835, an enterprising researcher by the name of Munk Af Rosenschöld had been doing experiments on electrical conduction in solids. And still earlier, in 1833, Michael Faraday was learning about temperature coefficients of solids.

These tidbits of knowledge kept adding to a growing mass of information about solids, yet they remained unrelated, waiting for the beginnings of electronics communications. Although wireless communication did start to stir during the latter portion of the 19th century, it wasn't until the end of World War I that the direct ancestor of the transistor made its appearance as the cat's whisker crystal. This was a true semiconductor, but its function was solely that of a rectifier. It is amusing to wonder whether, if the triode vacuum tube had not appeared to replace the crystal detector, some enterprising soul might not have found practical ways of making crystals amplify and oscillate. A few loyal investigators were sufficiently impressed with semiconductors to continue their work. Gabel V. Lossev in 1924 described the possibilities of the crystal as a generator and amplifier. And, in 1928, R. S. Ohl obtained a patent for a multi-electrode crystal that would amplify and oscillate. He was also the discoverer, in 1941, of the pn junction, basically the heart of the modern transistor.

But we must assign the formal birthdate of the transistor to June 1948, when Bell Telephone Laboratories first announced it. This transistor, a type not now used, was called the point contact and basically was a germanium diode, but with two cat's-whiskers instead of the one associated with the old-time crystal detector. Containing three leads, known as the emitter, collector, and base, the device could be used to amplify and oscillate.

Of course this was just the beginning. If the transistor was somewhat slow in getting started, compared to the vacuum tube, this tiny device has made gigantic strides. That brings us to the purpose of this revised edition. New developments in the field of solid-state electronics are growing by geometric progression! We begin here by explaining what the transistor is, how it works, and how it behaves in the presence of other components. We will then look at many different kinds of circuits that employ transistors. An entire chapter has been devoted to the field-effect transistor, in its myriad forms. We conclude by examining some practical transistor circuits, along with a construction project: a pocket-size shortwave receiving antenna.

The treatment here is not mathematically complex. Many explanations are given by analogy. Comprehension of the basic principles is more important, at the outset of any learning process, than total rigor. This book will give the beginning student of transistors a thorough, general understanding of this modern miracle. And, if this book whets your appetite for more detailed and rigorous knowledge, so much the better!

Chapter 1

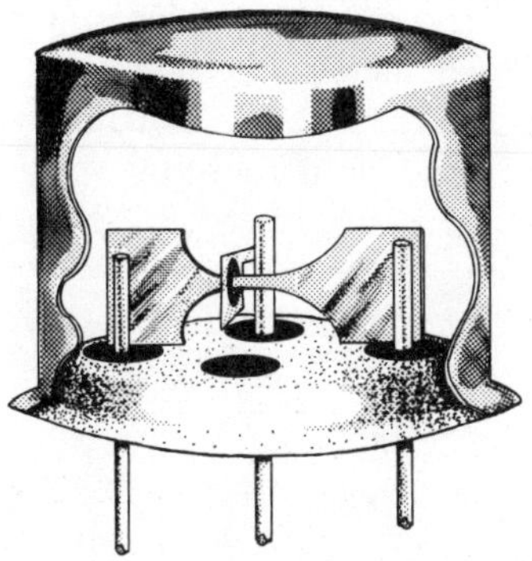

The World of the Atom

WE KNOW YOU EXPECT US TO TELL YOU ALL ABOUT TRANSISTORS (and we will) but let us start by taking a slight detour and talking about nails. What is a nail? Just a bit of iron, shaped in a special way and intended for a special purpose.

But when we say a nail is iron, we're indulging in what is known as a half-truth. Iron is a very sociable sort of metal and very rarely will you find it all by itself. For the most part, it likes to associate with oxygen, and when it does, we call the partnership iron oxide (rust).

This union of iron and oxygen is a very strange one. By itself, oxygen (at our usual temperatures) is a gas. We take in a lungful every time we inhale. And it is highly unlikely that you have ever seen pure iron, or that you would recognize it as iron, if you did. Pure iron is as white as silver, malleable and ductile. All this means is that it is so soft you wouldn't dream of using it to help build a house. But join iron and oxygen, add a pinch of carbon, and what a difference you get. You now have a strong, dark metal with which you can build houses or bridges.

ELEMENTS

What man hath joined together, man can put asunder. We can take iron oxide, separate the two partners, and have our pure iron (soft and silvery) and our oxygen (a gas) once again. Iron and oxygen belong to a family of substances called elements. Elements form a very exclusive society, since there are only about a hundred of them, but their importance is far out of proportion to their numbers, because they form the building blocks of the universe. Everything you wear, eat or drink, everything you see, including yourself and the people around you, is made of elements. Some you know very well or, at least, their names are familiar. You know arsenic, aluminum, calcium, chlorine, carbon, helium, hydrogen, and iron, but there are some that, until recently, were not so publicity-conscious. But we are going to stir up these very shy ones—germanium, selenium, silicon, indium, boron, and gallium—and give them the sense of importance they deserve.

COMPOUNDS AND MIXTURES

Elements, like people, have very distinctive characteristics. And, like people, they may change these characteristics when they get into a group or they may remain as individualistic as they ever were. Iron and oxygen unite to form iron oxide, a substance that is as different in its characteristics from iron and oxygen as it could possibly be. Sodium, a soft, silvery-white, waxlike metal, joins enthusiastically with chlorine (a nasty gas) to form sodium chloride. We know it much better as ordinary salt and we need it as part of our diet. Carbon, hydrogen and oxygen, having rid themselves of their inhibitions, group together to form alcohol (Fig. 1-1). Add some coloring matter, invite a few congenial people, and we have ourselves a party. But how different carbon (think of coal or diamonds), hydrogen and oxygen (both gases) are from alcohol.

Fig. 1-1. Alcohol is just one of the many thousands of compounds that can be formed from carbon, hydrogen, and oxygen.

This grouping, or combination, of elements to form something completely new, something with entirely different characteristics, is called a compound. Water is a compound, made of hydrogen and oxygen. Salt is a compound. And so is iron oxide.

Does this mean that the penalty for being sociable is loss of identity? Not always. We can take one compound, such as salt, shake it together thoroughly with another compound, such as pepper, but neither compound, peper or salt, will change its characteristics. The pepper remains sharp or tangy and the salt remains salty. A loose union of this sort in which the substances retain their individualities is called a mixture.

ATOMS

Now that we've moved somewhat in one direction, the joining of elements, let's make an about face and start marching off in the other. What happens when we start slicing elements apart, finer and finer and finer . . . ? The end product is an extremely tiny particle of the element, known as an atom. We say end product because, if we tamper with the atom in any way, it loses its identity. An atom of pure iron behaves in the same way as any other atom of pure iron, but start taking one of them apart and it is no longer an atom, and it is no longer iron.

Actually, though, the atom is as far as we are going to go. Taking the atom apart is the role of the atomic physicist. We are going to dally in that very pleasant, and instructive playground barely long enough to get some basic information about transistors.

Just about the simplest of all the atoms we could start with is the atom of the element hydrogen. Figure 1-2 is a drawing of what we think an atom might look like. Note that this atom has only two parts—a central body called the nucleus and a single, solitary electron revolving around it. This electron, in its motion around the nucleus, follows a definite orbit.

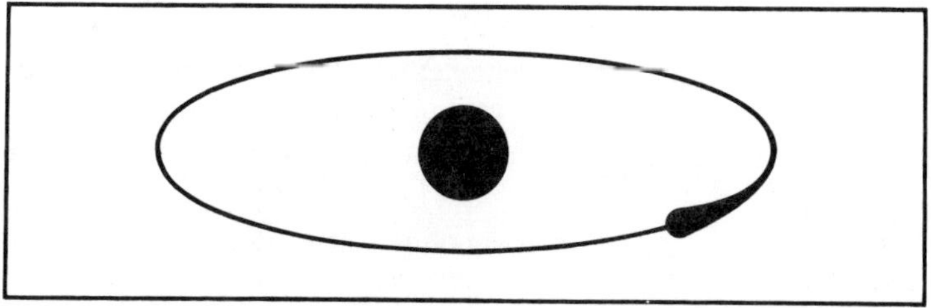

Fig. 1-2. The lonely world of the hydrogen atom. The nucleus has just a single orbital electron to keep it company.

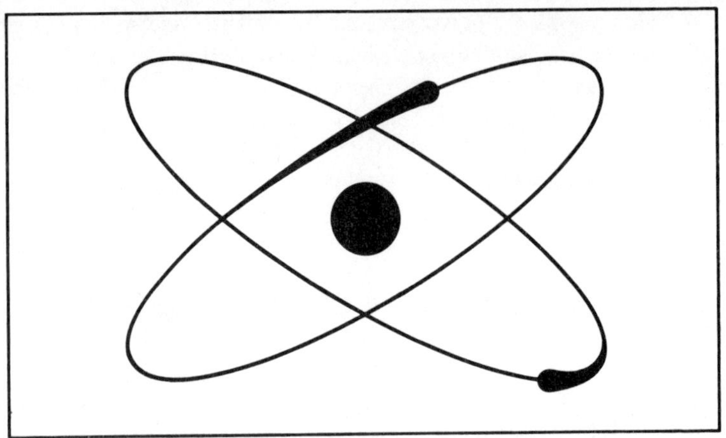

Fig. 1-3. The helium atom is content with its two orbital electrons. The helium atom is stable.

An atom is matter—mighty small, but it is matter, and as such it has mass. This applies both to the central part of the atom, the nucleus, and to its restless associate, the electron. Oddly enough, although the nucleus of the hydrogen atom has a mass that is about 1,845 times that of the electron[1], it is the electron that we are interested in. It is the diminutive member of this duo, the electron, that lets us have radio and television, that is responsible for modern communications.

There are a few things we know about the electron and there are quite a few things about the electron that have us guessing. We know it has a negative charge[2], and has a magnetic field around it. It is possible that the electron has angular momentum—that is, it spins or rotates around an axis. This knowledge of the vital statistics of the electron is singularly useful because it enables us to persuade the electron to leave its nice, comfortable orbit and to go traveling for our benefit. Specifically, we have transistors because we are able to make electrons do our bidding.

In the matter of electrons, hydrogen is the poorest element, since it has but one. The next element, helium, has two orbital electrons (Fig. 1-3). But now comes a sad story—a story of envy, desire and greed, the story of the haves and the have-nots. An atom

---

[1] The electron has a mass of 9.107×10^{-28} gram. The diameter of an electron is believed to be 2×10^{-12} centimeter. All electrons are identical, whether free or part of an atom.

[2] Each electron has a charge of 4.80×10^{-10} C.G.S. electrostatic units.

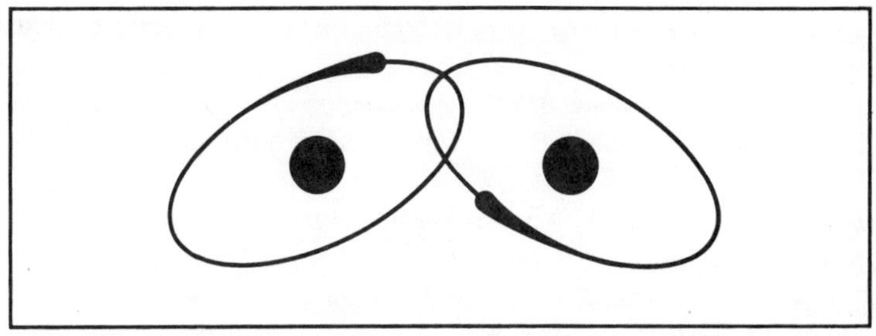

Fig. 1-4. Two hydrogen atoms can satisfy their yearning for stability by sharing electrons.

with two orbital electrons (and helium is an example) is a satisfied, contented atom. It has no ambition. It has no desire to acquire more electrons and, even if offered some, will reject them.

But what about hydrogen, poverty-stricken hydrogen, with its single electron? It schemes day and night to form a stable, happy family group, so reminiscent of helium. But, electrons are not easily come by, and so the hydrogen atom shrewdly does the next best thing—it joins forces with other hydrogen atoms, just as shown in Fig. 1-4. Each hydrogen atom shares its electron with its neighbor, and so, in that sense, we can say that the hydrogen atom has finally satisfied its yearning to have two electrons. But what about atoms (see Fig. 1-5) that have more than just one or two electrons?

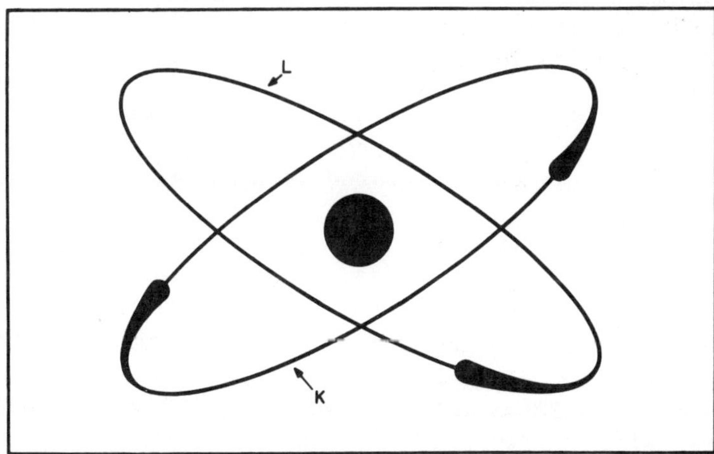

Fig. 1-5. The lithium atom has three electrons in two orbits. The two innermost electrons form an orbit or ring of their own. This is known as the K ring. The remaining electron has a separate orbit called the L ring. The single electron is much further removed from the large, centrally located nucleus, than the two electrons in the K ring.

We don't know why two electrons revolving around a nucleus makes the atom satisfied, but it does. One hint lies in the fact that the next atom, lithium, has three electrons. Two of lithium's electrons play in an orbit of their own which we have marked with the letter K in Fig. 1-5. But what about the third electron? It must revolve in an orbit all by itself. We have marked this orbit with the letter L.

Is there anything about the lithium atom that seems familiar? Doesn't it seem to you that there is some resemblance to our smug, self-satisfied helium atom (two electrons), and also to the dissatisfied hydrogen atom (with just one electron)? Actually, such is the case. The electrons in the K group are complete but, give it half a chance, and the lithium atom will associate with some other atom just so that its outermost electron (the single electron in the L ring) will have a playmate.

OTHER ELEMENTS

There are more than 100 different elements, each one with its own special properties. As you can see, adding just one electron and proton can make a big difference in the behavior of an element. Helium has characteristics greatly different from lithium, with one more electron and proton. Carbon bears little resemblance to nitrogen. Figure 1-6 shows us the first 11 elements pictorially. Table 1-1 lists all the known elements (at the time of this writing).

Electrons seem to want to arrange themselves into definite orbits. We call these orbits shells, because they surround the nucleus or kernel. The first shell, called the K shell, may have 2 electrons. If we try to add a third electron, it cannot go into the K shell, but must find another orbit, farther away from the nucleus. This orbit is called the L shell. As we continue to add more and more electrons to an atom, they all fall into the L shell until there are 8 in that shell and the whole atom has 10 (you'll recognize that as neon, a very stable gas). The poor little 11th electron gets a cool reception from the L shell; it is forced to find a new home, called the M shell. It certainly must feel lonely there, since the M shell has room for 18 electrons! When the M shell is full, the atom has 28 electrons; the unlucky 29th is cast out still further, into the N shell with 32 vacancies. The atom with 29 electrons is the element copper, an excellent conductor of electricity.

We can see that some atoms are "satisfied," that is, their shells are all filled to capacity. But there aren't many of these totally happy atoms in the world of the elements. Copper, with one lonely elec-

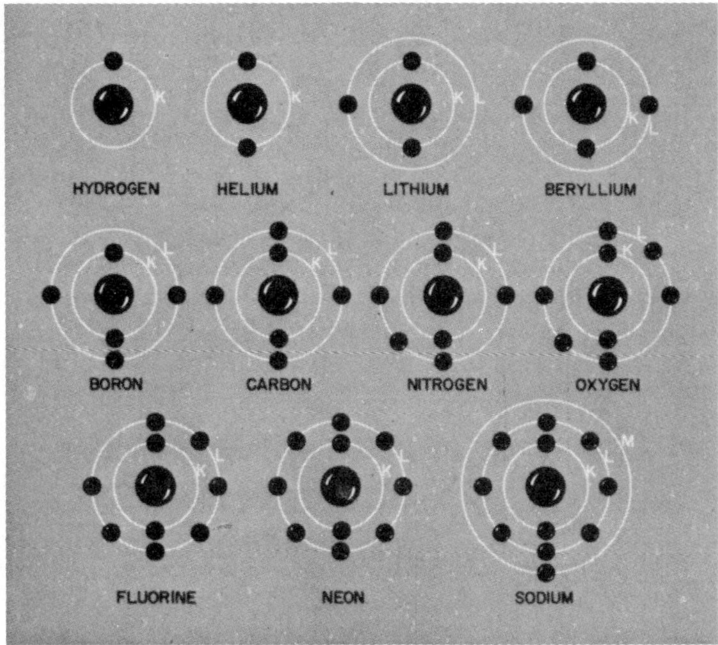

Fig. 1-6. One of the differences between atoms is the addition of electrons in the various rings. Hydrogen has one electron, helium has two, lithium three, etc.

tron far from the nucleus, is an extreme malcontent. Some atoms are just itching to throw their electrons around! These are the good electrical conductors, with an outer shell that's almost empty. A lonely electron travels often. Some atoms are in-between; they have partially filled shells. These atoms are not "happy," nor are they "malcontent." They are fair conductors, but not excellent ones like copper or silver. These are the semiconductors. It is they that make solid-state electronics possible.

PLEASE PASS THE ELECTRON

As you have probably noticed by now, we're spending quite a bit of time talking about electrons. This is exactly as it should be, since electrons are our stock in trade. We're not really interested in all electrons, just those we can persuade to leave home and travel. Like people, some electrons can be influenced to do what we want; others cannot. And since this is so, let's separate those who will from those who won't.

If a shell, or ring, of electrons is complete—that is, if the ring

Table 1-1. The Elements

Element & Symbol	Atomic Number	Outer Shell	Element & Symbol	Atomic Number	Outer Shell
Actinium (Ac)	89	2/98	Mendelevium (Md)	101	2/98
Aluminum (Al)	13	3/18	Mercury (Hg)	80	2/72
Americium (Am)	95	2/98	Molybdenum (Mo)	42	1/50
Antimony (Sb)	51	5/50	Neodymiun (Nd)	60	2/72
Argon (Ar)	18	8/18	Neon (Ne)	10	8/8
Arsenic (As)	33	5/32	Neptunium (Np)	93	2/98
Astatine (At)	85	7/72	Nickel (Ni)	28	2/32
Barium (Ba)	56	2/72	Niobium (Nb)	41	1/50
Berkelium (Bk)	97	2/98	Nitrogen (N)	7	5/8
Beryllium (Be)	4	2/8	Nobelium (No)	102	2/98
Bismuth (Bi)	83	5/72	Osmium (Os)	76	2/72
Boron (B)	5	3/8	Oxygen (O)	8	6/8
Bromine (Br)	35	5/32	Palladium (Pd)	46	18/32
Cadmium (Cd)	48	2/50	Phosphorus (P)	15	5/18
Calcium (Ca)	20	2/32	Platinum (Pt)	78	1/72
Californium (Cf)	98	2/98	Plutonium (Pu)	94	2/98
Carbon (C)	6	4/8	Polonium (Po)	84	6/72
Cerium (Ce)	58	2/72	Potassium (K)	19	1/32
Cesium (Cs)	55	1/72	Praseodymium (Pr)	59	2/72
Chlorine (Cl)	17	7/18	Promethium (Pm)	61	2/72
Chromium (Cr)	24	1/32	Proactinium (Pa)	91	2/98
Cobalt (Co)	27	2/32	Radium (Ra)	88	2/98
Copper (Cu)	29	1/32	Radon (Rn)	86	8/72
Curium (Cm)	96	2/98	Rhenium (Re)	75	2/72
Dysprosium (Dy)	66	2/72	Rhodium (Rh)	45	1/50
Einsteinium (Es)	99	2/98	Rubidium (Rb)	37	1/50
Erbium (Er)	68	2/72	Ruthenium (Ru)	44	1/50
Europium (Eu)	63	2/72	Samarium (Sm)	62	2/72
Fermium (Fm)	100	2/98	Scandium (Sc)	21	2/32
Fluorine (F)	9	7/8	Selenium (Se)	34	6/32

Francium (Fr)	87	1/98
Gadolinium (Gd)	64	2/72
Gallium (Ga)	31	3/32
Germanium (Ge)	32	4/32
Gold (Au)	79	1/72
Hafnium (Hf)	72	2/72
Helium (He)	2	2/2
Holmium (Ho)	67	2/72
Hydrogen (H)	1	1/2
Indium (In)	49	3/50
Iodine (I)	53	7/50
Iridium (Ir)	77	2/72
Iron (Fe)	26	2/32
Krypton (Kr)	36	8/32
Lanthanum (La)	57	2/72
Lawrencium (Lw)	103	2/98
Lead (Pb)	82	4/72
Lithium (Li)	3	1/8
Lutetium (Lu)	71	2/72
Magnesium (Mg)	12	2/18
Manganese (Mn)	25	2/32
Silicon (Si)	14	4/18
Silver (Ag)	47	1/50
Sodium (Na)	11	1/18
Strontium (Sr)	38	2/50
Sulfur (S)	16	6/18
Tantalum (Ta)	73	2/72
Technetium (Tc)	43	1/50
Tellurium (Te)	52	6/50
Terbium (Tb)	65	2/72
Thallium (Tl)	81	3/72
Thorium (Th)	90	2/98
Thulium (Tm)	69	2/72
Tin (Sn)	50	4/50
Titanium (Ti)	22	2/32
Tungsten (W)	74	2/72
Uranium (U)	92	2/98
Vanadium (V)	23	2/32
Xenon (Xe)	54	8/50
Ytterbium (Yb)	70	2/72
Yttrium (Y)	39	2/50
Zinc (Zn)	30	2/32
Zirconium (Zr)	40	2/50

There are presently 103 known elements. The atomic number is the number of electrons normally in a given atom. In the column labeled "outer shell," the number of electrons in the shell is given first, followed by the total possible number of electrons that may occupy that shell. Elements whose outer shells are filled are poor conductors. Elements with only one electron in the outer shell are good conductors. Other elements exhibit varying degrees of conductivity.

contains 2, 8, 18, or 32 electrons, don't even bother trying to talk them into leaving. These are the home bodies. We have a few of them in Fig. 1-6. Helium is one of this group. It has one ring (K) of 2 electrons, is a stand-patter, and that's that. The next one is neon with a total of 10 electrons. No chance of borrowing any electrons here, since the K ring is complete with 2 electrons and the L ring is filled with its quota of 8.

SOME ELECTRONS MOVE—SOME ELECTRONS WON'T

Now you can begin to appreciate why we started our study of semiconductors with atoms and why we did not plunge immediately into an explanation of transistors. We're interested in getting electrons to move when we tell them to. A small knowledge of atomic structure is helpful because we know in advance what elements will be cooperative and which elements will be stubborn.

Fortunately for us, very few atoms are as complete as helium or neon. Most of them have electrons that jump about almost as fast as women at a bargain hunt at a charity bazaar. Let's examine two of them, shown in Fig. 1-7. Here we have lithium and fluorine. Lithium's inner ring is 2 electrons, so we don't bother with it. But that outer electron has an itch to travel. Adjacent to the lithium atom, we have one of fluorine. Fluorine also has an inner ring of 2 electrons, but its outer ring has a juicy total of 7.

How do lithium and fluorine compare? Lithium is "electron poor." It has just 1 electron in its outer ring and not much opportunity to reach a grand total of 8. But what about fluorine? Its outer ring has almost made its quota. Just one more little electron and the L ring of fluorine is complete.

Now if we bring lithium and fluorine near each other, just what do you think will happen? Figure 1-7 gives us the result. The single, solitary, out-in-space electron of lithium moves over to fluorine. This union of lithium and fluorine gives us a molecule[3] which we call lithium fluoride.

We're not interested in teaching you chemistry—that is quite incidental to this book. The important thing to learn from Fig. 1-7 is that we now have one way of making an electron move. We don't say it is the best way (for us), but it does show us that electrons do get about.

Of even greater importance, Fig. 1-7 illustrates *which* electron moves. It is the electron in the outermost ring or orbit. And because

[3] A molecule is the union of 2 or more atoms, forming a compound.

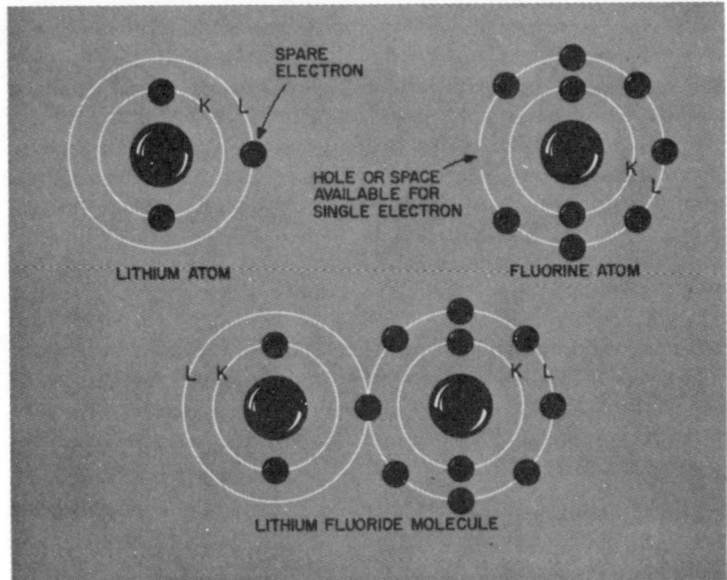

Fig. 1-7. Lithium has an electron it can spare. Fluorine has a vacant space into which this electron can fit. The two atoms, lithium and fluorine, form a molecule of lithium fluoride. This is a chemical combination. The molecule of lithium fluoride is a compound that has its own characteristics, completely different from that of either lithium or fluorine.

it is so important, so fundamental to our work in transistors, let us examine this union of atoms with another example.

Figure 1-8 is a drawing of a sodium atom. This atom has three rings, two of which we cannot touch. The first, or K ring has the usual 2 electrons, the next or L has a complete and very satisfying full quota of 8. But orbiting out in the space around the nucleus we have a single electron in the M shell.

In the same drawing we have pictured the chlorine atom. This also has three rings, two of which are filled. The M ring, though, has 7 electrons. And, as in the case of lithium and fluorine, the one electron in the outermost shell moves over. The result of this very fortuitous joining of sodium and chlorine is sodium chloride.

OTHER COMBINATIONS

We selected lithium and fluorine, sodium and chlorine, because they lent themselves so nicely to the idea of a moving electron. But other atoms join in just the same way. In each case, though, the only electrons we are interested in are those that can (and do) move, and these are *always* the electrons in the outermost orbit. This is true whether the atom has two shells or more than two. And because of this, we can simplify matters for ourselves in talking about other atoms by forgetting all electrons except the outermost ones.

ELECTRONS IN MOTION

Getting electrons to move is easy. To get the picture, we can compare their motion to the flow of water, but we must bear in mind that this is not a really correct description. In Fig. 1-9A we have two cans of water, one of which has a higher water level than the other. Connect the two with a pipe and water will flow until both cans have the same water level. Now let's try the same idea with electrons. In Fig. 1-9B we have a container and let us just imagine that we have somehow managed to fill it with electrons. Standing nearby is another container with far fewer electrons. And connecting the two is not a pipe, but a copper wire. Although we have no way of seeing directly what is going on, electrons will move from the electron-rich container to the one having not so many, until both have the same amount.

Now you might argue that it isn't possible to have two contain-

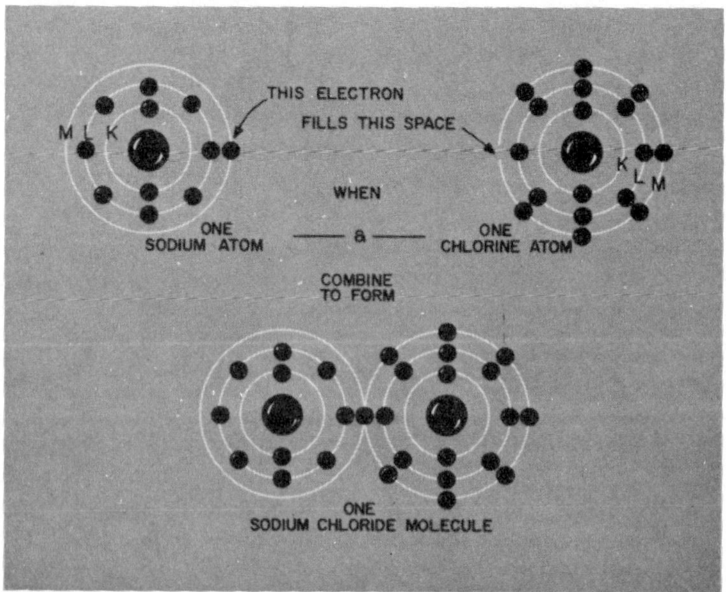

Fig. 1-8. The sodium atom has an electron it can spare in its outermost ring (M) while the chlorine atom has room for one more electron in its outermost ring (M). The union of these two atoms produces a molecule of sodium chloride.

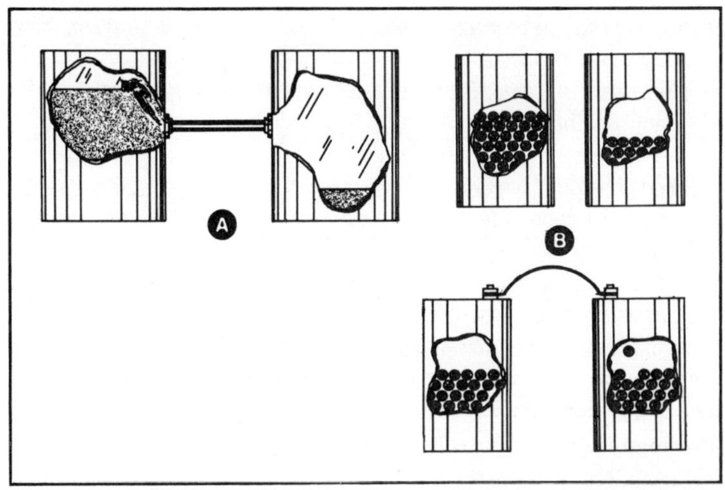

Fig. 1-9. In drawing A we have two containers, one of which has a higher water level than the other. If we connect a pipe between them, water will flow until both containers have the same quantity of water. In drawing B we have two containers, one of which has more electrons than the other. If we connect them with a wire, electrons will flow until both containers have the same number of electrons.

ers, one with, and the other without, electrons. But any battery presents almost that situation. A battery is a chemical factory, the end product of which is the condition whereby one electrode is loaded with electrons while the other has a terrible shortage of them. And if you wonder how it is possible for such a situation to come about, just remember that a battery is a very convenient arrangement for robbing some atoms of their outermost electrons. Some atoms are constantly losing electrons, some are always gaining them, and some (as in the case of neon or argon) never swap.

CONDUCTORS

The copper wire we connected between our two containers of electrons, or the terminals of our battery, is called a conductor because it supplied a passageway for electrons. Copper is another one of those elements whose atoms have very unsatisfied outer rings. Copper has four rings, 2, 8, 18 and 1 electron. This last ring, or shell, is so empty that it accepts all contributions with gratitude.

Now suppose we connect our copper wire to the negative terminal of a battery. Can't you just see the very first atom of copper saying to the electrons, "Come over and join my outermost ring"? And so they do. But the copper atom adjacent to the first one is also unsatisfied, so it demands, and gets, the electrons. This leaves the first atom shy of electrons again, but why worry? There are more available. And so the electrons are handed along, from atom to atom, much like buckets of water passed down the line by a 17th century fire brigade.

Now let's move over to the very last atom of copper, the one nearest the positive terminal of the battery. Do you think the last atom on the line is going to be allowed to hold on to its precious hoard of newly acquired electrons? Hardly! The positive terminal of the battery has even fewer electrons than this atom, and so it demands, and gets, its share.

When will this sort of thing cease? When the positive terminal of the battery has as many electrons as the negative terminal. Of course, at that time, there will no longer be such a thing as positive or negative terminals, since both will have the same number of

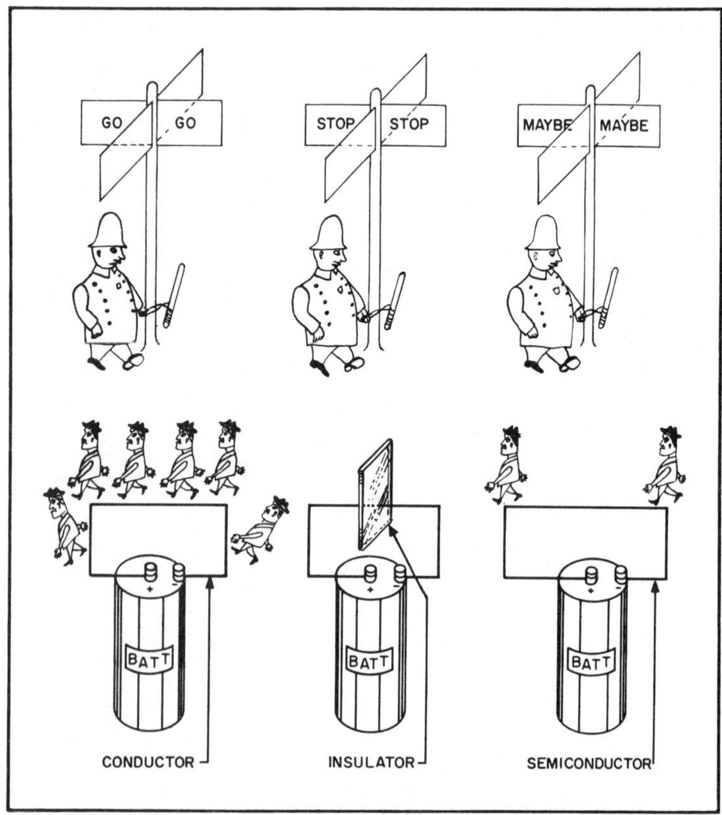

Fig. 1-10. A conductor permits the passage of electrons; an insulator does not. A semiconductor comes somewhere between these two extremes.

electrons. And what about our copper wire? Will it have been permitted to keep any of the electrons it was so busy passing along? Sorry. The battery gives electrons and the battery takes them away.

Our copper wire, willing and gullible, ready to work without electron compensation for any slick battery that comes along, is called a conductor. A conductor permits the easy passage of electrons through it (Fig. 1-10).

INSULATORS

Once a compound has acquired all the electrons it will ever need, any attempt we make to supply it with more will be rebuffed. Generally these are compounds whose outer rings have joined and so wouldn't know what to do with extra electrons if offered on a silver platter. Glass is one such compound. It has all the electrons it needs, thank you, so connecting a battery across it is a nice gesture, but a useless one.

SEMICONDUCTORS

Substances that are very good conductors or very good insulators are unique and represent a minority group. They belong to the all-or-nothing-at-all groups. We can take certain insulators, modify them somewhat to form a let's-straddle-the-fence-group, which we call semiconductors. It is from this group that we are going to build our transistors.

The trouble with oversimplification (which is what we tried to do in the paragraph above) is that it can lead you down the garden path to some incorrect thinking. An ideal insulator could be represented by an open circuit, or an infinite amount of resistance. An ideal conductor could be considered as a short circuit, or zero resistance. Between these two limits we have a tremendous number of resistors which are able to pass current to a greater or lesser degree. We can call them resistors or we can call them conductors, just as we prefer. They do *not*, however, include semiconductors. A semiconductor is not something that is halfway between an open circuit and a short circuit. As we mentioned in the preceding paragraph, a semiconductor starts out in life as an insulator and is then modified to give it certain characteristics.

BACK TO THE ELECTRON

Germanium, a grayish-white, brittle sort of metal is among the more important semiconductors. In some ways it is like carbon or silicon and, in others, like tin. Germanium is a mineral and, like almost all other minerals, has a crystalline structure. Salt is also crystalline. Look at it under a magnifying glass sometime and you will see how each face of this mineral has a definite geometric pattern. This doesn't mean that salt and germanium look alike, any more than two hats. Each mineral has its own definite structure.

Germanium has four electron shells, 2, 8, 18, and 4. It is the outer ring that fascinates us, since it is responsible for some rather unusual behavior. Actually, having four electrons in the outermost ring is what helps make germanium unique. It really doesn't fit in with those other atoms, like lithium, which loses electrons readily, or like fluorine, which obtains electrons so easily. Germanium is a sort of halfway point between atoms that surrender electrons and those that take them in. We have a drawing of a germanium atom in Fig. 1-11 and as you can see, the atom is sort of at loose ends with its incomplete outer shell.

But what will happen if we introduce our germanium atom, not to some stranger, but to some other germanium atom, also beset with the same problem. Which of the two will surrender its electrons? Which of the two will be the more demanding, and get electrons? What we have is a tug-of-war between equally matched contestants. Neither side wins and neither side loses.

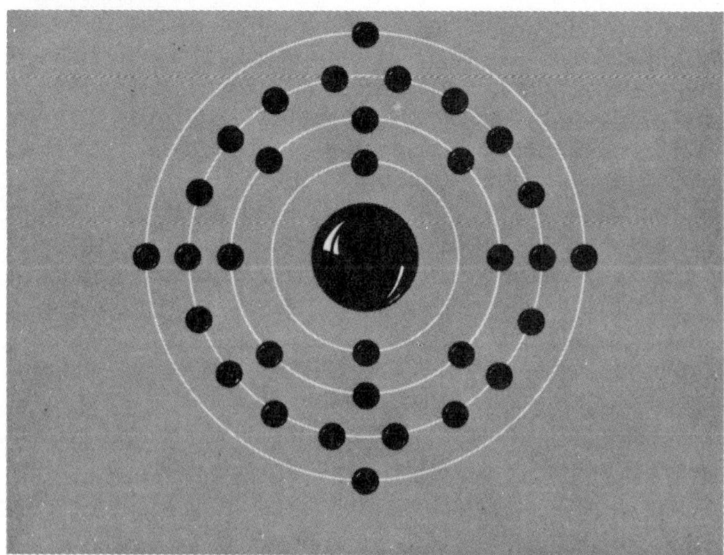

Fig. 1-11. The germanium atom has 4 electrons in its outermost ring.

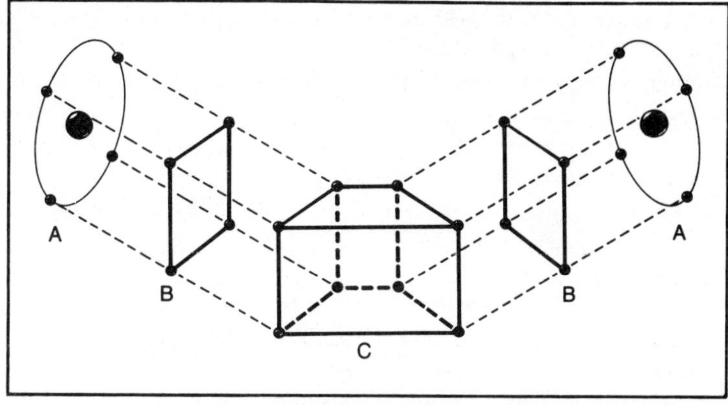

Fig. 1-12. In A we have a pair of germanium atoms. Just the outermost ring of electrons is shown. We can imagine these removed from the atoms, as in B. In C we have the 8 valence electrons, 4 from each germanium atom, forming a bond. This is atomic union, not chemical, as in the case of sodium and chlorine.

VALENCE

Up to now we have really been talking about two kinds of electron behavior. We have the stay-at-home, stick-in-the-mud types that form the ultraconservative complete rings or shells. And we have the let's-go-traveling-and-see-what-other-atoms-are-up-to types. But even here we must make some sort of distinction, since not all of the electrons in the outer incomplete rings are truly and equally venturesome. Let's go back to the case of lithium and fluorine. Lithium has 1 electron in its outermost orbit and fluorine has 7. Should they all get equal credit for having some get up and go? Not likely! There is just one electron that does any moving, so we should distinguish it in some way. We can, by giving it a new name. We can call it a valence electron.

But having endowed the lithium electron with this verbal croix de guerre, what shall we do about germanium? Figure 1-12 shows what happens. The electrons don't move from one germanium atom to the next, but they do sort of link hands. Technically speaking, each germanium atom now has a complete outer ring of 8 electrons. True, it gets this way by a sharing process, but the ring is still complete and that is just what the atom wants. While the electrons didn't move, they do deserve some sort of special recognition, so let us say that they have formed a valence bond.

BACK TO CONDUCTORS

A little earlier we connected a copper wire across a battery and saw how we could get an electron movement, or drift, through the wire. We were aided by the fact that the outermost rings of all the copper atoms were incomplete and were happy to get electrons from the battery, even if only on a highly temporary loan basis.

Now let us repeat our experiment, but this time let us make our wire out of germanium. We connect our wire to the battery and stand back expectantly, but nothing happens. Why? Our battery is fresh, its negative terminal is practically crawling with electrons and its positive terminal is suffering from an electron famine. Conditions are just ripe for the passage of electrons through the germanium wire—or are they? Why should the germanium wire pass electrons? What have the germanium atoms to gain? Absolutely nothing. Linked arm in arm by mutually shared electrons—by valence bonds—they are self-sufficient, and have no need of electrons. No current will flow and we have no choice but to classify germanium among the insulators. A rather odd situation, since germanium is a metal.

DOWN WITH THE VALENCE BONDS

Possibly you find the idea of germanium as an insulator slightly irritating. But if so, what can be done about it? One method (and we do not offer it as a practical technique) is to heat the germanium. As the temperature rises, some of the valence bonds weaken. Some of the outer rings assume an attitude of haughty aloofness. In this condition they have only 4 electrons and will welcome electrons from the battery. Later we will learn that this tidbit of knowledge is useful in protecting the transistor and keeping it, lemming-like, from attempted suicide.

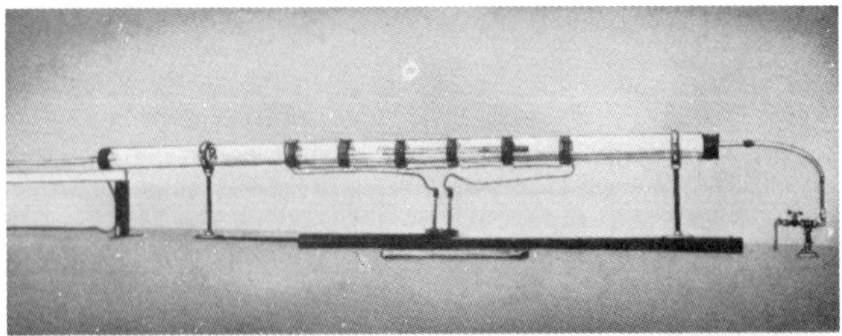

Fig. 1-13. Technique for refining germanium. This uses a method known as zone refining. The coils produce heated zones in succession, the impurities being pulled to one end of the tube. If the refining is repeated often enough, the amount of impurities remaining will be less than 1 part in 1 billion. (Bell Telephone Laboratories, Inc.)

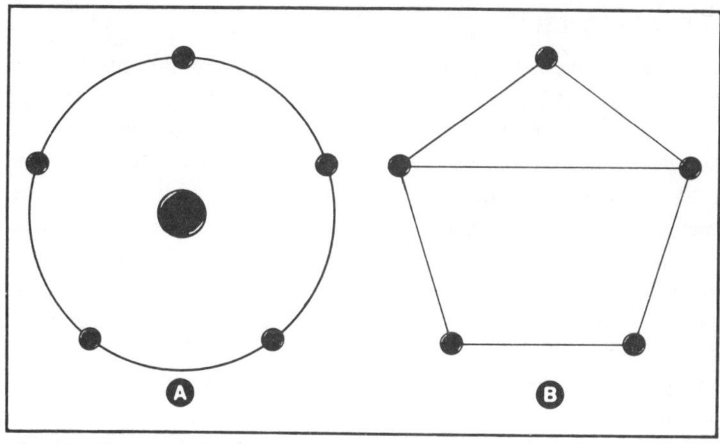

Fig. 1-14. The atom of antimony has 5 electrons in its outermost ring, as in A. The inner rings have been omitted to simplify the drawing. We can connect the electrons as shown in B. Antimony has 1 electron more in its outer ring than germanium.

ADD A PINCH OF THIS OR THAT AND STIR THOROUGHLY

The germanium we have been talking about so far is absolutely pure. You might argue that an absolutely pure element does not exist, but in the case of germanium at least, methods are known and used which make germanium 99.99999 (add a few more nines of your own) percent pure (Fig. 1-13).

But now, into this undefiled mass of germanium atoms, let us drop a tiny impurity. Did you ever see a kitchen floor that was absolutely spic and span, but with one tiny, muddy footstep? What did you notice—the whole expanse of clean floor, or the small spot of mud? We're not going to use mud for our germanium but we can add a different element to it, and because this element is different, we can call it an impurity. (The process of adding the impurity is known as doping.)

In Fig. 1-14 we have a drawing of the element we are going to use. This element, antimony, has no less than 5 rings, or shells, of electrons. The innermost has the usual 2 electrons, the next has 8, followed by two complete rings of 18 electrons each. And in the outermost shell we have 5 electrons.

As far as the outer shells are concerned, how does antimony compare with germanium? Germanium has 4. Antimony has 5. Do you remember our description of germanium? Antimony sounds almost like a twin, since it is tin-white in color, and is a hard, brittle metal having a crystalline structure.

LATTICE STRUCTURE

Before we stir up a witches' broth by adding antimony to germanium, let's go back for just a moment to germanium itself and consider an important fact about the construction of that element. Did you ever see a beehive? A marvel of geometric construction, isn't it? And if you could see the way germanium is put together, you would marvel at it too. Germanium has a lattice structure. Whenever we have a definite arrangement of atoms in a geometric pattern, we have a lattice. This is why germanium is crystalline. The most beautiful example of crystalline structure is the diamond. The surfaces of a diamond are arranged symmetrically, but the surfaces are just an indication of the extremely orderly arrangement of the atoms inside.

Atoms, like people, can come in clumps or bunches without any sort of order or plan. Think of people walking back and forth, and cross-wise along a pavement. Or, people can form orderly groups, as in a well organized parade.

ADDING THE ANTIMONY

Now antimony is a crystalline substance too and so, when we

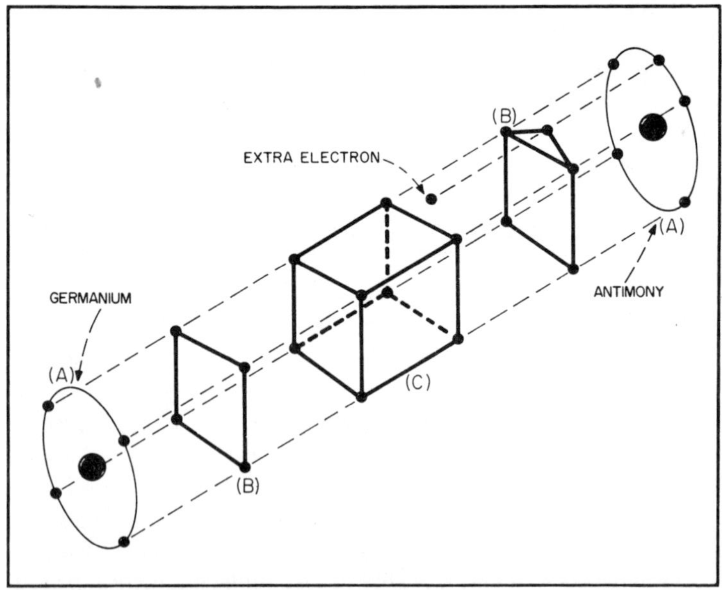

Fig. 1-15. When antimony is added to germanium, the result is a lattice structure—a bond between the germanium and antimony valence electrons. There is one electron, though, that isn't strongly bound.

put a tiny speck of it in with the germanium, we are adding one element with a lattice structure to another with a similar lattice structure. All well and good, but there's an electron in the ointment (Fig. 1-15).

If antimony had only four outermost electrons instead of five, all would be fine. Four electrons—yes, but that fifth electron is about as necessary as an extra guest when all the chairs are filled. But that extra electron, that orphan, should not be overlooked, for now it can serve a useful function. Now you might think that the antimony would want to keep its number 5 electron, but such is not the case. Its lattice structure fits right in with the germanium. It can form valence bonds with four of its outer electrons. What is the antimony doing ? It has gone native, hasn't it? It no longer needs or wants its surplus electron and is quite willing to release it and let it go elsewhere.

Now we can try our experiment once again, but this time let us use a wire made of germanium to which we have added a tiny amount of an impurity—antimony. We connect this wire across our battery. This time the electrons on the negative terminal push over onto the wire. The extra electrons of the antimony atoms, unwanted, are repelled by the electrons from the battery, and edge their way over to the positive terminal. But as soon as an electron leaves the outer ring another one hops in to take its place, only to be displaced by another electron moving forward toward it. But where have we heard this story before? Isn't this somewhat like the movement of electrons along the copper wire?

Our germanium is now no longer in the insulator class. At the same time, we really can't claim that it belongs with the good conductor group of which copper is such a prominent member. Since it is in a sort of limbo between the two, we call germanium, doped with an impurity such as antimony, a semiconductor.

OTHER IMPURITIES

The person who tracked up our nice clean kitchen floor a few paragraphs ago could also have brought in other impurities. Similarly, we need not depend on antimony alone. We could also use phosphorus or arsenic. Phosphorus has only three electron rings, but the outermost one has 5 electrons. Arsenic has four electron rings, but once again the outermost shell has 5 electrons. Unfortunately, we don't have an unlimited supply of elements with 5 electrons in the outer shell. Bismuth is the only other one. It has no less than six rings, but it still meets our requirements.

THE DONORS

These elements we have been talking about, antimony, arsenic, phosphorus and bismuth, always come bearing gifts. The gifts are the extra electrons so essential if we are to make germanium into a partial conductor. We call these elements donors, since they really do come equipped with a present of an electron. And since electrons are negative, we really should call them negative donors. Of course, the germanium, recipient of extra electrons, should be called negative germanium to distinguish it from germanium that has retained its essential purity. And negative germanium it is, abbreviated as n-type germanium.

The union of germanium and a donor element such as antimony or arsenic isn't at all like the joining of lithium and fluorine we talked about earlier. Lithium and fluorine combine chemically to form a completely new compound. An example with which you are more familiar is the formation of salt (sodium chloride) from sodium and chlorine. The salt doesn't look at all like its forbears, doesn't act like them, and is the true chemical union called a compound. But let's get back to germanium and the antimony. We add some antimony to the germanium but no new compound is formed. The germanium remains what it is and so does the antimony. The intermingling of the two is on an atomic basis. The nearest analagous situation we have in practice is putting butter on hot pancakes. The butter melts and mixes with the pancake until every tiny available space is filled with it. But the pancake remains a pancake and the butter is still butter.

THE ACCEPTORS

Did you ever see the number 4 written as 4 ± 1? What this means is that we can have 4 plus 1 (to give us 5) or 4 minus 1 (resulting in 3). What has this to do with us? We've been talking about germanium atoms with 4 outer shell electrons and about donor elements all of which have 5 outer shell electrons. In other words, 4 plus 1. This gives us a clue to the idea that we might intermingle another type of element with our germanium. This element would have only 3 electrons in its outer shell.

What elements could these be? Boron is one of them. Boron has just two shells, a complete K shell having the usual 2 electrons, and an outside shell of 3 electrons. Aluminum is another one but has three rings; 2 electrons for the first, 8 electrons for the second and 3 outermost electrons. We have to travel down a long list of elements to come to the next one—gallium, with four electron rings: 2, 8, 18, and 3. The next is indium with five rings: 2, 8, 18, 18 and 3. Then we have thallium with six rings—2, 8, 18, 32, 18, and 3. And that is all. There are no more elements with 3 electrons in the outside shell, but we have enough now, and any one of those we have mentioned is suitable for our purpose.

Now let us *diffuse* some boron into pure germanium. Did you notice that we italicized a word? We did so deliberately because the process of getting some boron into the germanium isn't just an ordinary mixing operation. The effect is shown in Fig. 1-16.

Of course, the boron atom, being part of a minority group, would like to do anything and everything to make people think that it is not a boron, but a germanium atom. But to be able to carry out this deception it needs just one electron, which it promptly borrows from its nearest neighboring germanium atom. But this germanium atom, having lost an outer electron to the predatory boron atom, promptly borrows an electron from the nearest adjacent germanium atom. And so it goes. Not a single atom wants to be short that single electron in its outer ring, and so we have a sort of aimless wandering of electrons from one atom to the next.

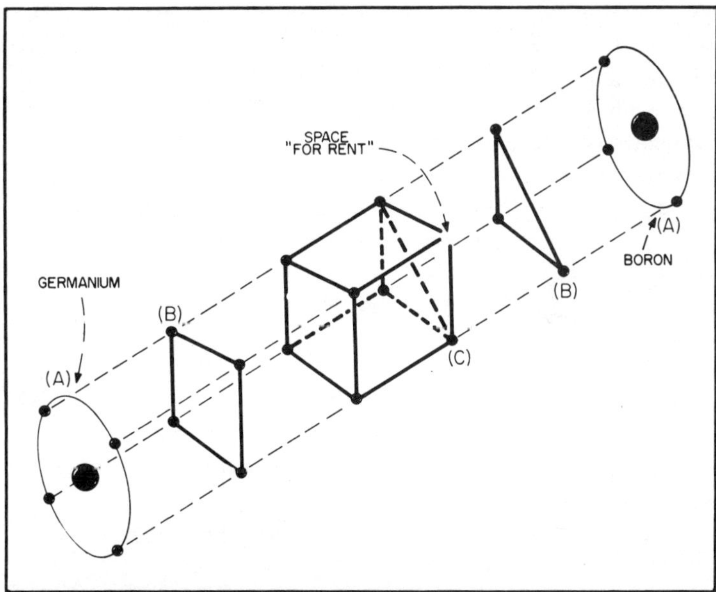

Fig. 1-16. Germanium and boron (drawing A) showing valence electrons. (Inner rings have been omitted for simplicity.) We can see the lattice structure by connecting lines as in B. But in drawing C we note that the crystal is incomplete. Boron does not have enough valence electrons and so we are left with a space or "hole" into which some electron could fit very nicely.

POSITIVE AND NEGATIVE CHARGES

Earlier we mentioned that electrons are not only particles of matter (but extremely tiny) but that they also carry a negative charge. Of course, the more electrons you manage to collect into one spot, the more negative that spot is. Suppose we do have such a location and the four electrons are sitting there, just as nice and comfortable as you please. And now, while we are using our imagination, let us also suppose that one of the electrons gets into a huff and departs. Our location now has fewer electrons. But precisely because it has fewer electrons, it is not as negative as it was before. And because it is less negative, we can say that it is more positive. It's just like saying that if a person becomes less rich he also becomes more poor. We know you may think this is just a play on words, but it is important for us to realize that we can say one thing in two different ways and to have both of them mean the same.

HOLES

Let us now consider the case of the unfortunate germanium atom that finds an eager-beaver boron atom located right next to it. Before the germanium atom realizes what has happened, it loses one of its four outer-shell electrons to the boron atom. We know that the germanium is going to steal an electron from its neighbor, in turn, but before it gets a chance to do so, let us "freeze" the picture, so that we can examine the germanium atom as it exists during the time of its loss.

The germanium atom has lost an electron and, since electrons are negative, the germanium atom is now less negative than it was before. But let's define a little more clearly just what we mean by less negative. If your shoe is too tight, it may be just one toe that hurts. You don't ache all over. And when a germanium atom loses an electron, it doesn't become positive all over. We're concerned only with the outer ring, and even there, not with the entire ring. Consider the region vacated by the electron. It is now positive due to the departure of the electron. The loss of the electron has left a space. We call that space a hole, and to add further to what you must surely think is a state of confusion, some people refer to it as a positive hole.[4]

Do you remember our description of a germanium atom? One of its characteristics is that it has 4 — not 3 — not 2 — not 5 — electrons in its outermost ring. For a germanium atom to really be a

[4] This is a redundancy. A hole is positive.

germanium atom, it must have its particular structure, and no other. Remove any part of the germanium atom, such as an electron, and you have made a change, an important change. It is no longer really and truly a germanium atom, since its architecture has now been made different. It is an atom with a positive charge.

But now let us take pity on our electron-deprived germanium atom and supply it with an electron—a negative charge. Where do you think that electron will go? There is only one place it can fit and that is in the "hole" created by the departure of an electron. If you want, we can say that the hole and the electron *recombine*.

A STRANGE GAME OF CHECKERS

We want you to imagine that we have a checkerboard in front of us and that we have covered every available space on the board with checkers. This is extremely unusual, but so are transistors, so we'll make up our own rules as we go along. We also want you to imagine that each checker has a big, bold minus sign painted right on it where we can see it. Each checker represents an electron, or a negative charge.

What is our first move? Actually, there is no move we can make since every space is filled. Doesn't this remind you of pure germanium, where every electron is in its place? There's just no room or opportunity for electron movement whatsoever.

Now back to our checkerboard again. Let's lift one checker, somewhere near the center of the board. For the most part this doesn't seem to help very much (Fig. 1-17) but we do have the ability now to move electrons. But before we do any moving, what about the space where the single solitary checker used to be? We could give it any number of names. We could call it a space, an enclosed area, a hole. But no matter what we call it we do know what it represents. We removed a minus charge , so the space or hole, or whatever we want to call it, is positive.[5]

Now let's start moving checkers. If we're patient about it we can make that hole or space move anywhere on the board we want. We can chase the hole progressively from one end of the board to the other. But as we do a very peculiar circumstance emerges. As the hole moves in one direction, electrons move in exactly the opposite direction. We can call the movement of the hole a move-

[5] The electron not only has a charge, but mass as well. It can exist independently of the atom. A hole, though, is a charge only. It depends for its existence on the existence of the atom. Holes, or positive charges, do not have a life apart from the atom.

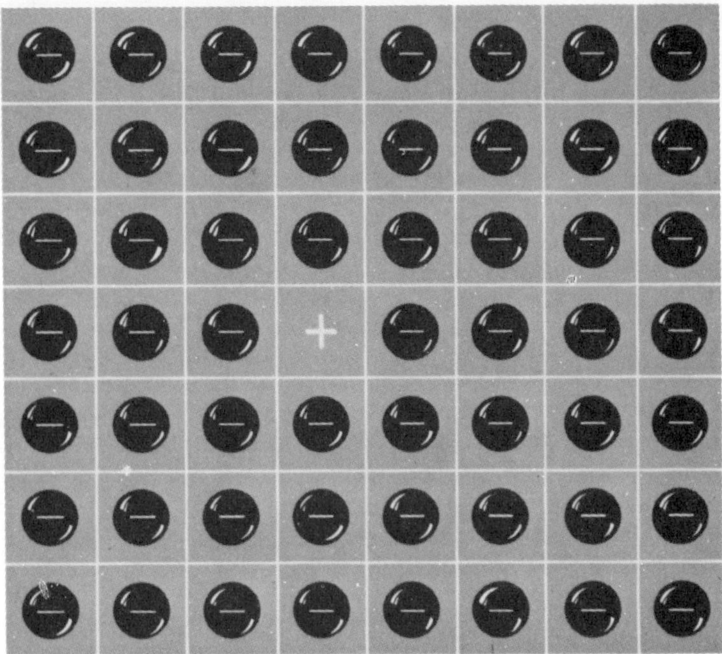

Fig. 1-17. All we need is a single vacant space on the checkerboard, and every electron on the board can be made to move. We can visualize this movement in two ways. We can think of the electrons moving about on the board, or, equally important, we can think of the space or "hole" taking different positions on the board.

ment of positive charge or hole flow. Or, if this represents too great a strain on our imagination, we can still think of electron flow.

We've been away from our germanium for quite a while now, so let's go back to it. Into a bit of pure germanium let's diffuse some boron. There's going to be a bit of an electron scramble as the germanium and boron atoms start their mad interchange of electrons. But, as in our strange checker game, there's always going to be a positive hole somewhere. Actually, we are going to have a tremendous number of holes because just a speck of boron provides millions of atoms. This is going to be a checker game on a really grand scale. We are going to have numerous positive charges, so it's perfectly proper and permissible to call this type of germanium positive germanium. Let's do some abbreviating and call it p-type germanium.

We now have two types: n-type germanium and p-type. Which we get depends entirely on the element we add to our pure germanium.

CURRENT FLOW

When we made a wire of n-type germanium and connected it across a battery, we had a trickle of current flowing, but there was nothing particularly unusual about it. But what about p-type germanium? What will happen when we make a wire of such germanium and connect it across a battery?

As usual, the negative terminal of our battery is crowded with untold millions of electrons all of whom are casting longing eyes at the wide open spaces on the positive terminal, so near and yet so far away. When we connect our p-type germanium wire, the electrons on the negative terminal can see but one thing . . . holes — positive charges — just willing and waiting to be filled by electrons. And so the electrons go hopping through the wire, moving along from atom to atom. And while the electrons move through the wire in one direction, we get the effect of the positive holes moving in exactly the opposite direction. If you have any trouble visualizing this, just look at Fig. 1-18 where we have a simplification of what is happening in our wire. As electrons move from left to right, the empty space or hole moves from right to left.

A STRANGE KIND OF BOOKKEEPING

A little while ago we played an unusual sort of checker game,

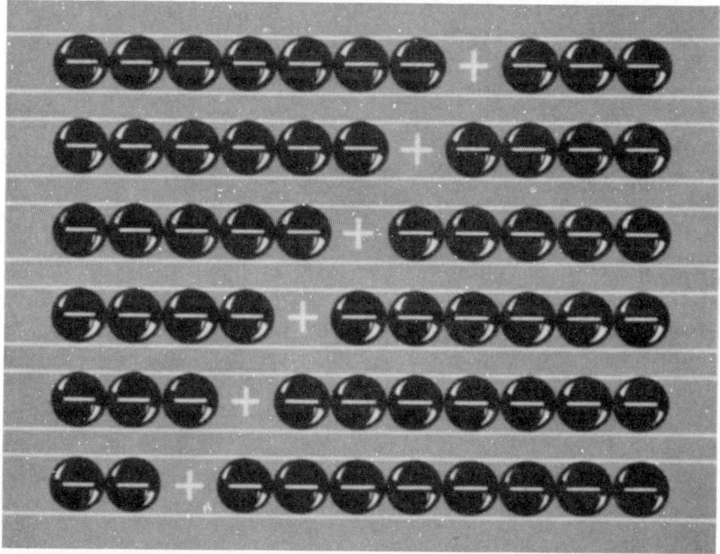

Fig. 1-18. As the electrons (negative charges) move to the right, the hole (absence of electron, or positive charge) moves to the left.

and now we are going to have an accounting system that is going to appear just as odd.

To understand just what it is we are about to thrust at you, please take a look at Fig. 1-19. We are still using our long-suffering battery, but this time we are going to ask you to imagine that we are able to insert an ammeter right in the battery itself, between the positive and negative electrodes. We also have an ammeter right at the negative terminal and another one at the positive terminal. Each of these meters is a triplet—that is, each one is absolutely identical. With our wire connected, each meter will read exactly the same.

Is this so surprising? Hardly! We have a series circuit and so the current must be the same in each and every section of that circuit. We are sure this isn't front-page news as far as you are concerned, but there is a moral and a lesson to be learned here. For every electron that leaves the negative terminal of the battery an electron must enter the positive terminal. And because of this, our copper wire never changes. When we disconnect it, after electrons

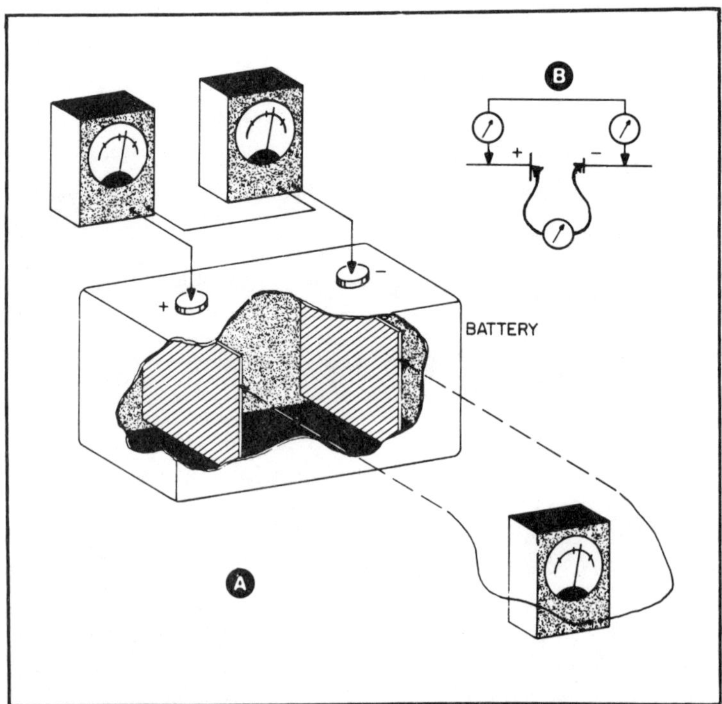

Fig. 1-19. A perfect bookkeeping system! Electrons are never lost. The same number of electrons that leave the battery return to it.

have passed through it from the battery, it is precisely the same as it was when we first connected it. The books are balanced. There has been no loss of electrons, nor has there been any gain.

Doesn't this apply to our germanium as well? Obviously a series circuit is a series circuit no matter what components we use, and equally obviously, the battery isn't going to change the way it operates just because we hang some germanium wire across its terminals. Again, what does this really mean to us? Simply this: p- or n-type germanium, like any other conductor, will let electrons enter from a battery, but the total number coming in will equal the same number going out. The germanium is not left with a surplus of electrons. All that the germanium has done, and all that it can possibly do, is to act as a passageway. The germanium doesn't care which element we use as an impurity either.

CURRENT MOVEMENT

There are just a few miscellaneous facts we should discuss before we go into the next chapter. We've been talking about wire made of germanium simply to convey an idea. It is doubtful if there is such a commodity and, if by some very strange coincidence there should be, it would certainly be lots more expensive than copper.

We've also been talking about the flow of current through copper and through germanium. The words *flow of current* may, unfortunately, create a completely wrong impression. The movement of current through a copper wire is not a mad dash. Far from it. It is much more like a leisurely drift, but it certainly does give the impression of being instantaneous.

Consider a circular tube filled with metal balls, just like a ball bearing, from which the idea was taken.

Now let's push one of the balls—any one. What happens? All the balls move. The effect of your push seems to have been transmitted almost without loss of time, from the first ball to the last. Actually, though, the ball you pushed traveled a very small distance only and so did the ball next to it, and the ball next to that one.

The movement of current through a semiconductor is at an even slower pace than through a conductor, such as copper. In the case of p-type germanium we assumed that the holes would form an absolutely straight line so that the electrons could go hopping, also in a straight line, from one end of the germanium wire to the other. No such thing happens and we have no right to make such an assumption. The current might actually spread out and take a rather devious route. We've also enthused about the beauty of the orderly lattice structure of germanium. But nature isn't perfect by any means, and so some germanium crystals may not be as well constructed as others. Bluntly, there may be imperfections and because of these imperfections the movement of the electrons may not be the smooth passage we imagine. Our meter in our series circuit won't reveal this to us, since it indicates just the average flow.

OTHER SEMICONDUCTORS

We've been talking so steadily about germanium that we have perhaps created the false impression that this is the only semiconductor. The requirement is simple and any element that meets the requirement is eligible for joining the semiconductor club. There aren't very many, however. Next to germanium, and recently of increasing importance, is silicon. Silicon has three electron rings (compared to four for germanium). The inner ring is the usual 2-electron orbit, the next ring has 8 electrons and the outer ring (the one that interests us) has 4. Although much of our discussion so far has been concerned with germanium, silicon is used more frequently in many semiconductor devices today because of its physical ruggedness.

Oddly enough, we have a somewhat unexpected member of the semiconductor family—lead. The ore of lead is lead sulphide or galena, and in the early days before vacuum tubes it was used as a semiconductor in crystal sets. We called them crystal sets since the galena looked like—and was—and is a crystalline substance. And so, if you think semiconductors are new, the answer is that they have been with us since the very early days of radio.

Lead has no less than six electron rings: 2, 8, 18, 32, 18, 4. If you can think of all these electrons whirling in orbit around the lead nucleus, your imagination is in fine working order.

Gallium arsenide is another semiconductor material. Like lead sulphide, it is a compound. Gallium arsenide is being used in the manufacture of tunnel diodes. Although we call it a diode (Fig. 1-20), it can work as an oscillator, it can amplify, and is very useful as an electronic switch.

Gallium arsenide uses materials we have come to regard as acceptor and donor impurities. Gallium has 3 electrons in its outer most orbit; arsenic has 5. A semiconductor made of these elements is known as a III-V compound. Another type of semiconductor along these lines is gallium phosphide. This is also a III-V crystalline compound semiconductor. Unlike germanium, some of these

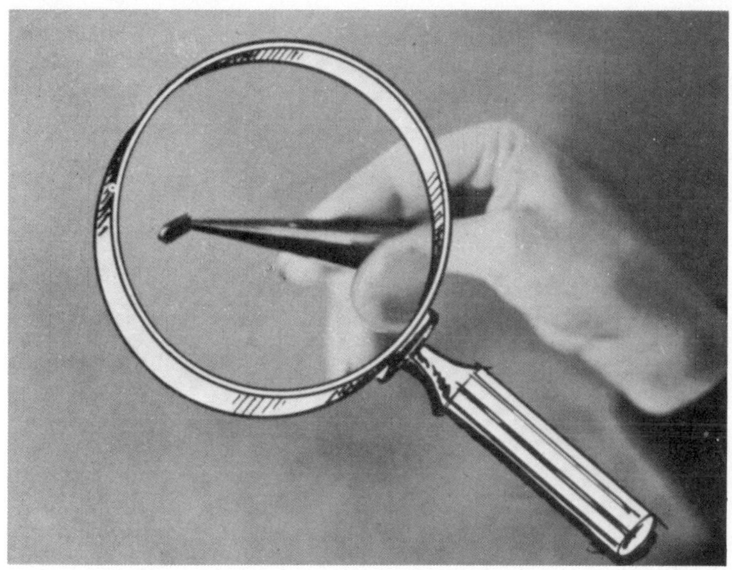

Fig. 1-20. Tunnel diode is so small that tweezers are helpful in handling it.

semiconductors will operate continuously at temperatures as high as 930°F. Thus, diodes produced from this material show promise for high-temperature operation, and as electroluminescent structures. The semiconductor itself is optically transparent. This makes it very useful in the basic study of semiconductors in general. For instance, it may actually be possible to see the differences caused by various degrees of doping and also the effects of various voltages.

The last of our semiconductors is tin, also rather surprising. But surprising or not, tin meets all the requirements. Tin is a sort of blue-white crystalline metal having five electron rings: 2, 8, 18, 18 and 4.

LOOKING FORWARD

The man who first realized that a slice of ham and a few slices of bread had unsuspected possibilities was a genious, for he invented the ham sandwich. He's in the same class with those who looked at p-type germanium, at n-type germanium and then combined the two, for this led directly to the invention of the transistor. How this was done and what happens when we make this sort of electronic sandwich will give us mental food for thought in the next chapter.

QUESTIONS

1. What is an electron shell? How many electrons make a complete K shell?

2. Describe what is meant by n-type germanium.

3. What is a donor element? Name two and describe the effect of their diffusion into pure germanium.

4. What is meant by p-type germanium?

5. What is a lattice?

6. Can a copper wire or a bit of germanium accumulate electrons when connected to a source of voltage? Explain your answer.

7. Describe a conductor. An insulator. A semiconductor.

8. How does the movement of electrons through germanium compare to the movement of electrons through copper? Is there a difference?

9. Describe the formation of positive charges in germanium.

10. What is the direction of hole movement compared to electron movement?

11. What is a valence electron? How does it differ from other electrons?

12. What is a valence bond?

13. What is an acceptor element ? Name two.

14. What difference, if any, exists between donor and acceptor elements?

15. What name or names do we use for the central portion of an atom?

Chapter 2

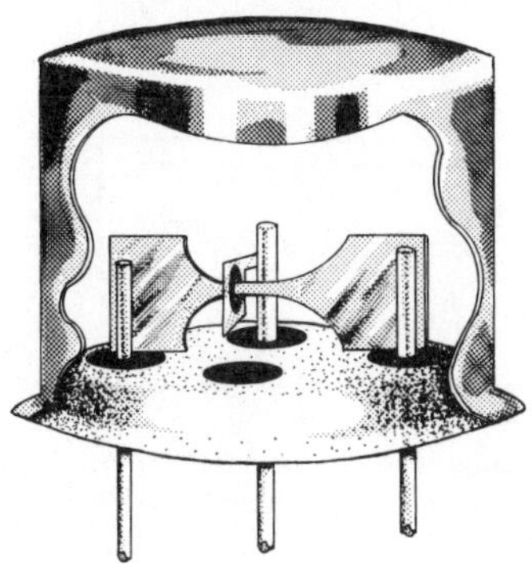

The Diode

BEFORE AN AUTOMOBILE CAN BE BUILT, ALL THE MATERIALS for its manufacture must be assembled. Since the same sort of thinking applies to the transistor, you now have one more clue to the purpose of the preceding chapter. We were getting the transistor's component parts all ready. These consist of n-type germanium and p-type germanium or silicon. And, just so that we can identify these items readily, we have drawings of n-type and p-type material in Fig. 2-1. These drawings, in their own exaggerated way, emphasize what we have already learned—that p-type material contains numerous atoms that suffer from electron shortages. Because of this, p-type material is shown as a rectangle with numerous plus signs scattered throughout. The atoms in the n-type material suffer from an embarrassment of richness—they have a surfeit of electrons and so we show this by numerous minus signs inside the little rectangle. More often, though, a semiconductor is shown in block form, as in Fig. 2-2.

Now you might very well object to the drawing in Fig. 2-1, since it may be difficult for you to regard the absence of an electron as a positive charge. But if you feel this way, consider a storage battery. One terminal is positive, while the other is negative. They are even stamped this way or have some other sort of identifying mark. But what is the true difference between the two terminals? The only real difference is that one terminal (the negative terminal) has far more electrons than the other terminal (the positive ter-

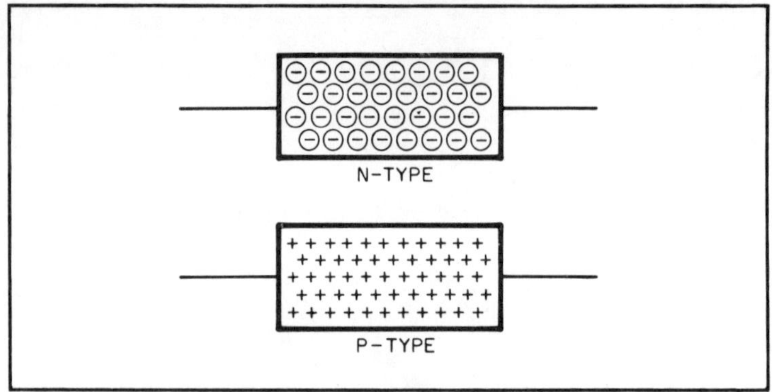

Fig. 2-1. We can represent n-type germanium (or any other semiconductor) by a rectangle filled with minus signs. These indicate an excess of electrons. For p-type germanium (or any other semiconductor) we can show a shortage of electrons by plus signs.

minal). Note precisely what it is we are saying here. The terms negative and positive are relative. We do not say the positive terminal has no electrons: just that it has fewer electrons than the negative terminal.

RECOMBINATION AND MOVEMENT

Have you ever been in a train that was packed so tightly with people that no one could move? We hope you have, because the inconvenience you suffered will now help you one step forward in understanding transistors.

Imagine a train, with no seats at all. The train is absolutely jammed with people, packed so tightly that no one can possibly move. The train pulls into a station and just one person gets off. This

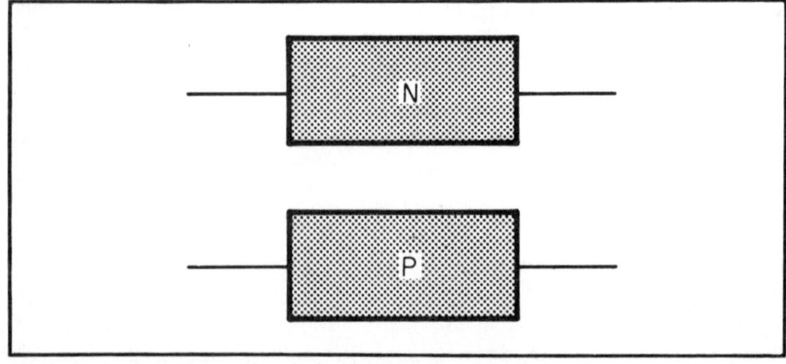

Fig. 2-2. A common method of representing n-type and p-type semiconductors in block form. These are not electronic symbols.

creates a vacancy (or hole) into which a person could move. And so now we have a condition in which some movement inside the train is possible. Thus, one person adjacent to the vacant space could move into it; but in so doing, he would also create a vacant space into which the next person could move. The important thing to notice here is that movement is possible.

Now suppose that by this time the train pulls into the next station and one person gets on. This person occupies the only available space, and so movement stops entirely.

What sort of conclusion does this bring us to? Simply that if a person is to move, there must be a space for him (or her) to move into. This isn't a revolutionary idea, but what happens to our thinking if we apply it to an electron? An electron has mass and it occupies space. But here the analogy between people and electrons breaks down. An electron has a charge. When an atom loses an electron—or gains one—the net charge of the atom increases or decreases (Fig. 2-3). The atom becomes more negative if it gains an electron, more positive if it loses one. And if an atom reverts to its status quo, it is neutral once again.

GRAND UNION

The ancient Chinese had a belief that the world was made of

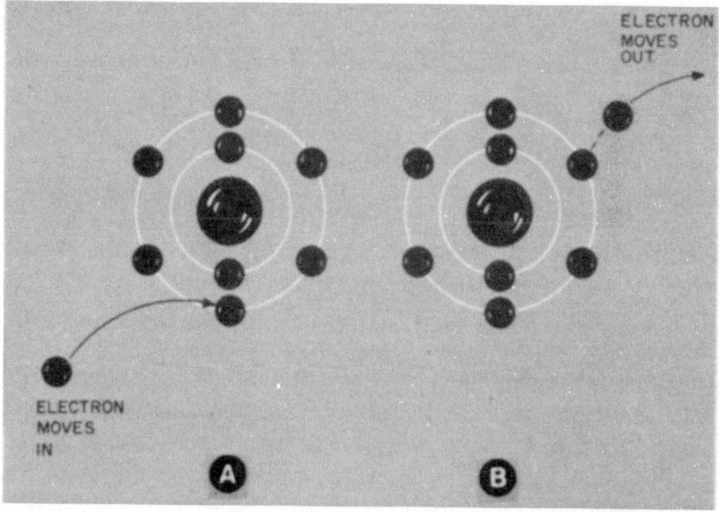

Fig. 2-3. An atom that has its regular quota of electrons is electrically neutral. It becomes an ion if it loses or gains an electron. If it gains an electron, it becomes a negative ion, as in drawing A. If it loses an electron (as in B), it becomes a positive ion.

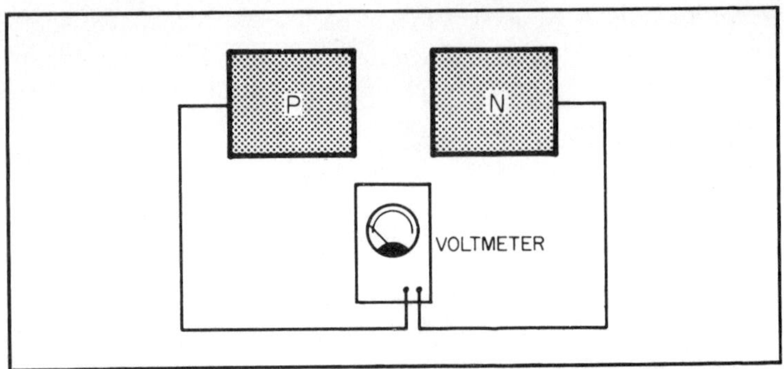

Fig. 2-4. The situation here is not the same as a battery. There is no movement of electrons from n-type to p-type material.

opposites, and these they called yin and yang. Many are immediately apparent—black and white, hot and cold, up and down, male and female, rich and poor. So it wouldn't surprise our friends of the Far East if we came up with n-type and p-type germanium or silicon. And because opposites have a habit of wanting to cling to each other, let's join our two types of semiconductor and examine the results. Before we do though, suppose we consider our two types just a bit further.

Our n-type material is called that because it is electron rich. Unfortunately, though, this sort of language may create a wrong impression. Figure 2-4 illustrates the problem. Here we have a block of n-type material and a block of p-type. We have a voltmeter connected across the two, but the voltmeter reads zero. Why should this be so? Isn't the situation here exactly the same as a battery? Don't we have more electrons on one side than the other? To get the answer, think back to the way in which we made n-type material. We added donor atoms. In trying to adapt its own structure to the lattice structure of the semiconductor, the donor atom creates a condition in which an electron has been released so that it can travel. That is all that has been done. The net charge of the slab of the n-type material is zero. Similarly with the p-type. We have added an acceptor element with the result that the p-type material is in a position to allow the passage of electrons. That is all we have done and the net charge of the p-type material is also zero.

But now let's put our two blocks of semiconductor in as intimate contact as we can get them. In doing so we create a very pleasant situation for the material. The single-valence electrons of the donors in the n-type material will now move because there is good reason for them to do so. Not too far away from them is an area of p-type material with holes just waiting and receptive to any stray electrons. And so we get a recombination of electrons and holes.

FAREWELL TO NEUTRALITY

The events that just took place did so at the junction of the two blocks of semiconductor material (Fig. 2-5). Our first thought might be that all of the spare electrons in the n-type material and all the holes in the p-type would combine, but they are prevented from doing so. Let's consider the atoms at the junction. What has happened to the atoms of p-type material? The holes of these atoms (at the junction) have disappeared. They've been filled with electrons. But because of this these atoms are no longer neutral. The atoms have acquired an electron and so are negative. Similarly the atoms of the n-type material (at the junction) have lost electrons and so are positive. These charged atoms at the junction form a barrier. This barrier is an electric field—or voltage, since we have had an electron movement.

Now we can see why all of the atoms in the p- and n-material do not swap charges. They can't. The barrier, or electric field, prevents the rest of the holes in the p-type material and the remainder of the electrons in the n-type material from getting together.

THE ELECTRIC FIELD

We cannot dismiss the condition existing at the junction of n- and p-type material blocks so summarily. The reason we may not is

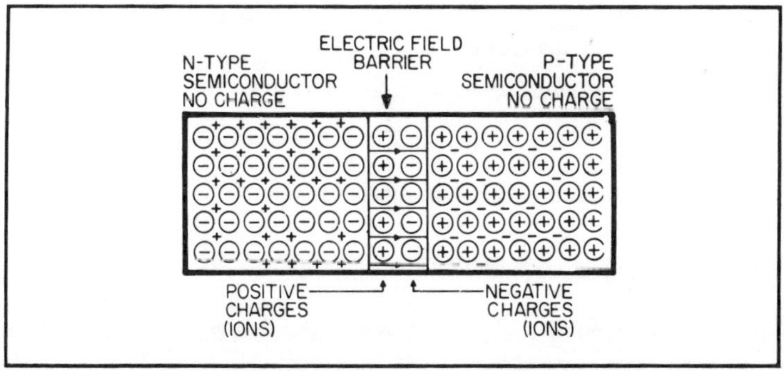

Fig. 2-5. The area at which n-type and p-type material (or any other semiconductor) meet is known as a junction. In this area we get a diffusion of electrons and holes. The result is a buildup of positive and negative charges which stops any further diffusion. The charges form an electric field, or barrier, which prevents any additional movement of charges.

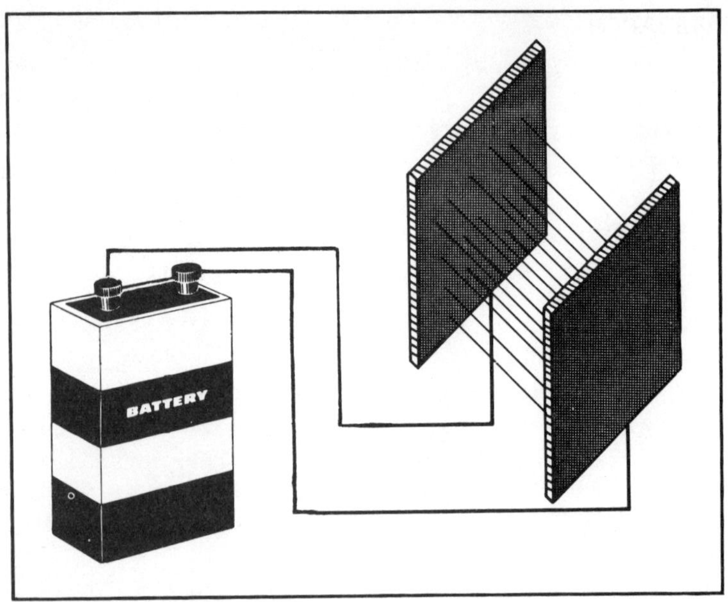

Fig. 2-6. This drawing shows a pair of plates connected to the positive and negative terminals of a battery. Electric lines of force exist between the plates.

that transistors work the way they do because of the barrier that exists. The barrier is the heart of the transistor.

Suppose, as shown in Fig. 2-6, we have a pair of metal plates connected to a battery. A force exists between the two plates. We cannot see this force any more than we can magnetic lines of force, but it exists just the same. And just as we can draw imaginary lines to represent a magnetic field, so too can we draw lines between the two plates. We can call these lines an electric field.

If we were able to pick up an electron and deposit it in this electric field, the electron (carrying a negative charge) would promptly accelerate over to the positive plate. And if you could, without overstraining your imagination, pick up a positive charge and put it in the region of this electric field, it would just as promptly move to the negative plate.

But let's be contrary for a while. Suppose we did put an electron into the electric field and decided that we wanted it to go to the negative plate. The electron would oppose our efforts. For the electron the negatively charged plate is a force in opposition—a sort of barrier. And we would have exactly the same results if we tried to push a positive charge toward the positive plate. Again we would get the same frustrated feeling of opposition—a barrier.

IONIZATION

It is very easy to see, in Fig. 2-6, how we get our electric field. It is supplied by the battery. A battery is something tangible. We can buy it, we can touch it, we can connect it. But how do we get an electric field at the interfaces (the junction) of our two blocks of semiconductor material?

To get the answer to this question we must think about our atoms once again. Suppose we deprive an atom of an electron. An electron, remember, carries a negative charge. So our atom, having lost a negative charge, is now considered positive—and quite rightly so. All we are really saying is that the atom isn't quite so negative as it was before the robbery took place. If our atom manages to acquire an extra electron we think of it as being negative.

Back in Chapter 1 we told you that atoms that gain or lose electrons are no longer the same. They have changed, and because they are different than they were originally, we call them ions. An ion is an atom that has gained or lost an electron. An ion is a charged atom and can be either positive or negative. An atom is neutral, but when it becomes an ion, that fence-straddling neutrality is gone.

BACK TO THE BARRIER

Armed with this information, let us see how an electric field is set up at the junction. Electrons migrate from the n-type material. They are attracted to those beautiful open spaces which are really positive charges. But the departure of an electron from an atom changes it from an atom to an ion. All along the interface of the n-type material we have ions. But what sort of ions? Since these atoms have lost electrons they must be positive, or plus, ions. What a strange thing to have happened. Our n-type material now has one side or face filled with positive charges.

What about the p-type material? The atoms along its face are gaining electrons. But if they are, they are no longer really atoms. They are ions—and negative ions at that!

And so, facing each other (Fig. 2-5) we have positive and negative ions—or, more simply, positive and negative charges. But how different is this from the charged plates in Fig. 2-6. If, in Fig. 2-6, we put a voltmeter across the charged plates, we would read a voltage. But isn't the situation the same in Fig. 2-7? In the case of Fig. 2-7, though, we did not need to depend on a battery to supply an electric field. Because of the ionization that takes place at the junction of the n- and p-type material, we have a measurable

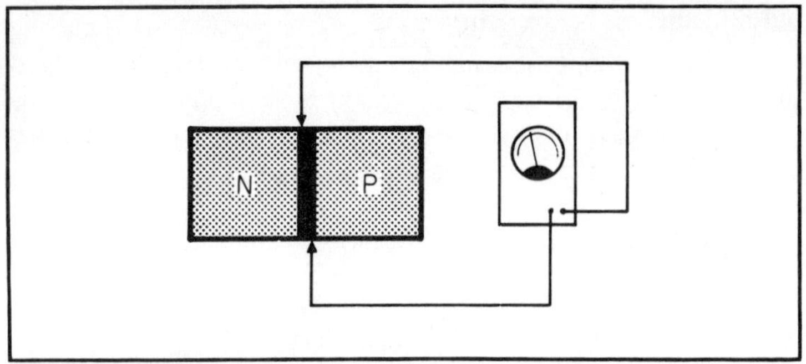

Fig. 2-7. A small, but measurable, voltage exists at the junction of the n- and p-type semiconductors.

voltage. The effect is just as though we had somehow managed to insert a tiny cell between the two semiconductor blocks.

Going back to Fig. 2-5, we have a drawing which sums up what we have done thus far. We have two blocks of material—p and n—closely joined. Where they join, we have the formation of ions. On the p-side, the ions are negative. On the n-side, the ions are positive.

What about the electrons in the n-type material? How do they feel about all this? They would like to get across to the p-type material, but they are unable to get past the positive ions. We have the same sad situation in the p-type material. Here the atoms would like nothing better than to play host to some electrons, but the negatively charged ions on their side stand in the way.

We now have an impasse. We have a potential, or voltage, across the interfaces of the two types of material. This is our barrier potential. The electrons in the n-type material need some help at this point since, on their own, they just do not have enough energy to overcome the potential barrier. At this stage of the game, we have a pn junction. This junction has its own electric field which keeps electrons on the n-side and holes on the p-side from getting together.

A MATTER OF POTENTIAL

We now have a rather odd situation. Here are two blocks of material—one n- and the other p-type. Each is electrically neutral. Yet some of the extra electrons in the n-type have moved across into the p-type. At the interface, the n-type, having lost electrons is now *positive*. And the p-type, having gained electrons, is now *negative*. This difference of potential, though, just exists at the interface.

AWAY WITH THE STATUS QUO

We have been talking quite a bit about the potential barrier because it is so important to our story, but compared to the total number of available electrons in the n-type material, the amount that actually move are trifling. If we want to transfer any substantial amount of electrons, we're going to need to supply the electrons with energy. We could do it with heat, but there is a more practical method available.

Figure 2-8 shows the technique we are going to use. Here we have connected a battery so that the positive terminal is attached to the p-type material and the negative terminal to the n-type material. Electrons will now move from the negative terminal of the battery into the n-type. What is this motion? It isn't a calm or leisurely affair. The battery must overcome opposition, actually force or inject electrons into the n-type material. It's like getting a shot for the flu. However cooperative the patient, and however sharp the needle, some amount of force is needed to accomplish the injection. In the case of silicon or germanium, the battery supplies the force, and the higher the voltage, the more force we have available.

What have we accomplished? We've injected electrons into the n-type material. But isn't this the semiconductor that, just a short time ago, lost some electrons at its interface? We even said that this region had become positive because of this loss. Now, we have re-supplied electrons to make up for that loss, so the number of positive ions at the n-side of the interface decreases. If we reduce the number of ions, though, what happens to our potential barrier? Since the potential barrier exists because of ionization, a reduction in the number of ions lowers the potential barrier. Wasn't it the potential barrier that kept more than a limited number of electrons from moving across the interface? With the lowering of the barrier, electrons can now move across from the n- to the p-type material.

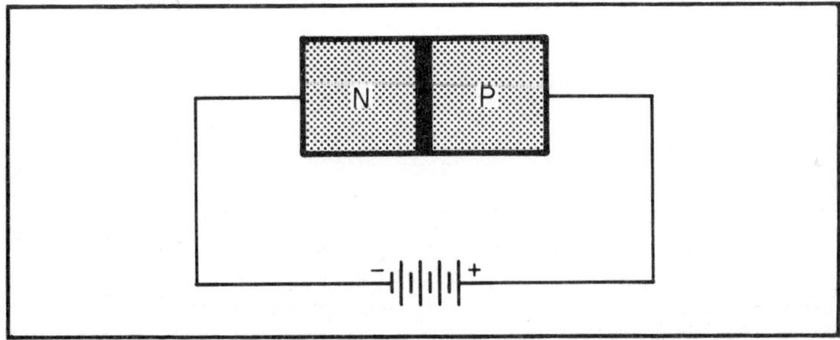

Fig. 2-8. We can use a battery to force current through the semiconductor.

WHAT'S GOING ON HERE?

When we first joined n- and p-type material, ionization, caused by movement of holes and electrons, could not exist until the two blocks of material were physically joined. When we finally did unite the material we obtained ionization on both sides. The voltage across the barrier—the barrier potential—soon put a stop to further movement of any charges. But if we lower the barrier potential, we lower the hindrance.

Does this mean that we eliminate the barrier completely? Not at all. We have lowered the potential difference across the interface but we haven't done away with it completely, since we will have ions constantly being formed. The situation has changed because we are now able, with the help of the battery, to de-ionize many of the ions being formed. Most—not all—of the atoms that lose electrons (hence become positive ions) will receive electrons because of that electron storehouse—the battery (Fig. 2-9).

Up to now, though, we have been neglecting the positive terminal of the battery and the bit of p-type material to which we have connected it. The positive terminal of the battery is characterized by a very important feature. It does have electrons—but far, far fewer than the negative terminal of the battery. What effect will this have on the p-type material? Actually, we should have been speaking about the ions at the interface. What sort of ions are these at the p-type interface? In this case we are dealing with negative ions—atoms that have one electron more than in their neutral state.

These negative ions now have a wonderful chance to become neutral once again. Their extra electrons, heeding the siren call of the positive terminal of the battery, soon start to move in that

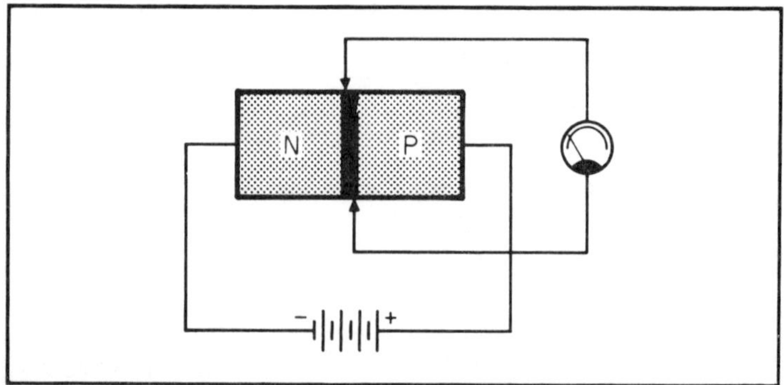

Fig. 2-9. The application of a voltage to the n- and p-type semiconductor reduces the potential barrier.

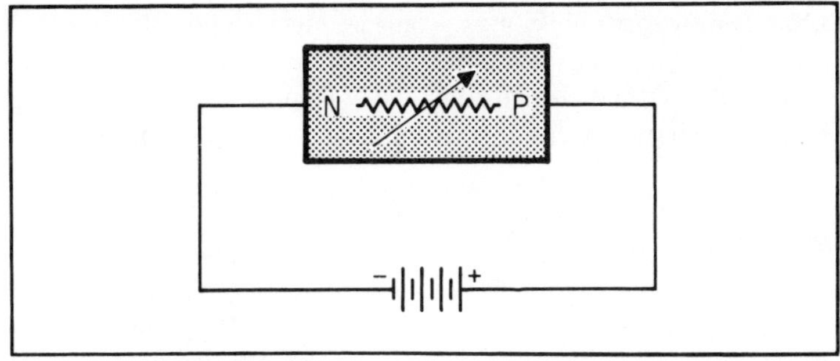

Fig. 2-10. The potential barrier acts like a variable resistor. Its value at any moment depends on the battery and the diffusion taking place at the interfaces of the n- and p-type semiconductors.

direction. Again we have de-ionization, only this time it is the negative ions that become neutral.

Let's recapitulate what happens when we connect our battery. The positive ions at the n-type interface acquire electrons and so de-ionize. The negative ions at the p-type interface lose electrons and so de-ionize. With de-ionization taking place on both sides, the potential barrier is lowered.

What happens to the electrons from an overall point of view? They move from the negative terminal of the battery, through the n-type material across the lowered barrier, through the p-type material over to the positive terminal of the battery.

How many electrons will flow? That depends on the potential barrier. If it is low, more current will flow. If it is high, less current. We can, if you like, regard the barrier as a resistance (Fig. 2-10). With a comparatively large potential barrier, we'll get small current flow. A large potential barrier would correspond to a high value of resistance.

HOLE MOVEMENT AND ELECTRON MOVEMENT

How shall we regard the movement of charges in our diode? We have three choices. We can consider it from the viewpoint of electron flow. Or we can think of it as hole flow. Actually, it is both, but there is a practical aspect we must consider. As long as we confine our attention to the semiconductor, and to the semiconductor only, the true action is a movement of electrons and holes, and in opposite directions. But the transistor is simply a means to an end. It will be used, just as a vacuum tube is used, to control current (electron) flow. Thus, in all circuits external to the transistor, in the wiring and the various components we are going to connect to the transistor, we are only going to consider electron flow, for that is all that will take place. Recombination—the reunion of holes and electrons, hole and electron movement combined, takes place only in the transistor. Outside the transistor we have only electron movement.

WHY, WHY, WHY?

It might seem that what we have accomplished thus far is not too important, but consider what has been done. We took silicon or germanium, which in their pure state are insulators, and made conductors out of them. But to do this required that we trick the material in some way. We did this by making one block of it so that it had an electron surplus, and another block so that it had an electron deficiency. When we joined the two blocks we did get a small amount of current drift, and if electron movement had not been blocked by ionization, we would have had a gradual diffusion, a gradual equalization of electrons on both sides. That equalization is never reached, since, when we connected a battery to the material, electrons were coaxed into entering the positive terminal of the battery as fast as electrons departed from the negative terminal.

FORWARD BIAS

We have so many different sizes, shapes, styles, and chemical arrangements in batteries that the word battery, used by itself, is somewhat meaningless. If we say flashlight battery, auto battery, or filament battery, then the modifying word adds some sense; it paints a picture. The battery that we used in Fig. 2-9 also needs a name. It is a bias battery. And because this battery helps overcome the potential barrier, because it makes numerous electrons march where relatively few moved, we call it forward bias. Note well that in forward bias the negative terminal of the battery connects to the n-block of the semiconductor; the positive terminal to the p-block.

REVERSE BIAS

The opposition to the flow of current, that is, to hole movement and electron movement, that exists as a result of the potential barrier can be lowered by using forward bias. But the control of current flow is not really complete if all we can do is to increase it.

To see how we can raise the opposition to current flow, look at Fig. 2-11. We are still using our two blocks of p- and n-type material. They are still in very intimate contact. The only change

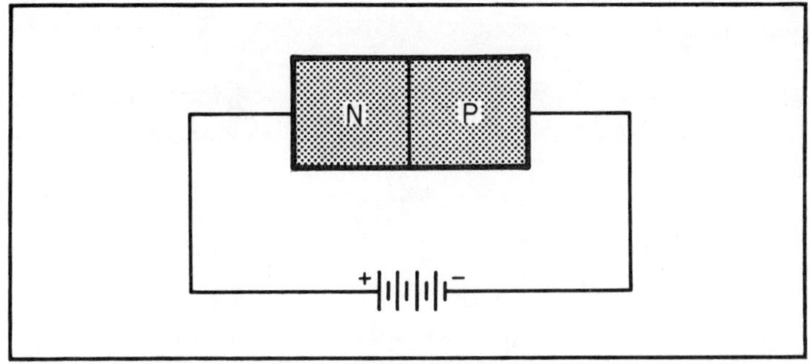

Fig. 2-11. Reverse biasing of a semiconductor diode.

we seem to have made is to transpose the battery. But this reversal of polarity produces important changes at the junction of the two semiconductor blocks.

The drawing in Fig. 2-12 shows that the number of ions on both sides of the junction has increased. We have more negative ions and we have an equal amount of positive ions. But the formation of ions on both sides results in a potential difference. This potential difference, or voltage, is our barrier, but now our barrier is stronger than it ever was.

What has produced this condition? The positive terminal of the battery, connected to the n-type material, has provided a haven for electrons. Some of these move into the positive terminal of the battery. At the same time an equal amount of electrons move way from the negative terminal of the battery into the p-type material.

In Fig. 2-13 we have a drawing in which we can compare the

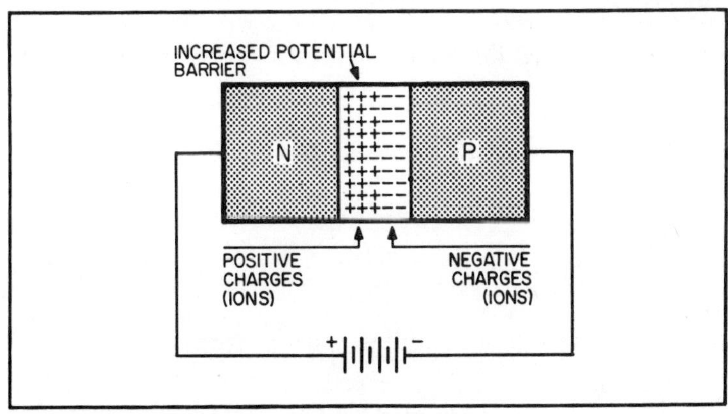

Fig. 2-12. Reverse biasing increases the potential barrier.

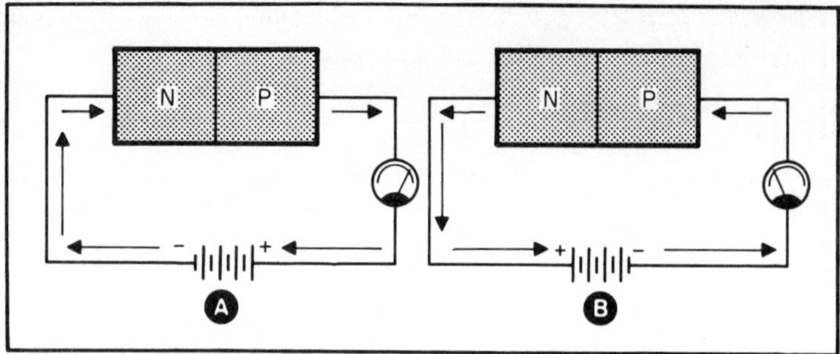

Fig. 2-13. The drawing in A shows forward biasing; the one in B is reverse biasing. Although current flows through the diodes in opposite directions its movement is still from minus to plus.

two ways in which we connected a battery to the semiconductor. In both cases the electron current flows away from the negative terminal of the battery, toward its positive terminal. And yet, as you can see by the arrows, the currents flow in opposite directions through the junction. To emphasize the opposite flow of current, a semiconductor diode that has a reverse (or opposite) flow of current is said to be reverse biased.

A MATTER OF DEGREE

We can get forward current or reverse current, depending entirely on how we connect our battery to the semiconductor diode. But there is an implication of equality here that is not true. Reverse current is small, generally measured in terms of microamperes. Forward current is large and is usually milliamperes or amperes. But you can no more consider current by itself than you can operate a seesaw alone. High current connotes low resistance. Similarly, low current means that its opposite, resistance, is way up in the air. Forward biasing is just another way of saying low potential barrier or low resistance at the junction. Reverse biasing goes hand in hand with high resistance at the junction or a high potential barrier.

THE DIODE

It may seem to you a little late in this chapter for us to give a name to our two blocks of semiconductor material. It is known as a diode, a rather lazy selection, since the name was borrowed directly from its vacuum-tube counterpart. The dictionary definition is certainly not applicable. The prefix is di (from the Greek *dis*, meaning two) while the suffix is ode (from the Greek *hodos*, meaning a way or path). Thus, a diode has two paths, and if we think of the forward and reverse flows of current, then perhaps the name diode isn't too poor a selection after all.

GRAPHS

Now you might wonder why we have spent so much time on the diode, since our basic interest is the transistor. There are a number of good reasons for this. The transistor (as we shall learn in our next chapter) is developed directly from the diode. To understand how the transistor works you must first know what happens in the diode. Actually (again, as we shall learn later) the transistor can be considered as a pair of diodes, back to back, so you are really studying transistors, possibly without realizing it. Finally, transistors can be used as diodes.

In Fig. 2-14 we have the graph of the forward and reverse currents of a semiconductor diode. Let's examine the graph to see what information we can extract from it. First, consider the situation in which the diode is forward biased. As the voltage is gradually increased the forward current increases. Now consider the case

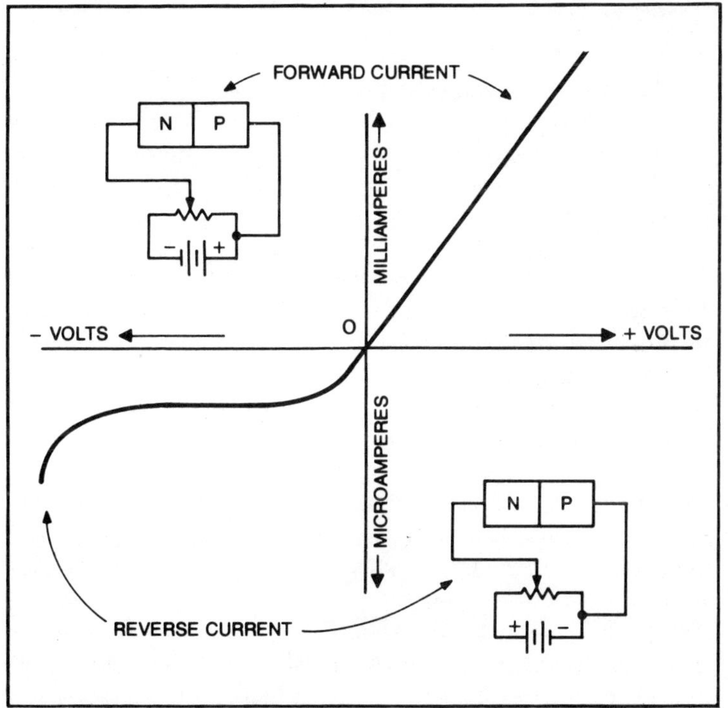

Fig. 2-14. Forward and reverse currents of a semiconductor diode.

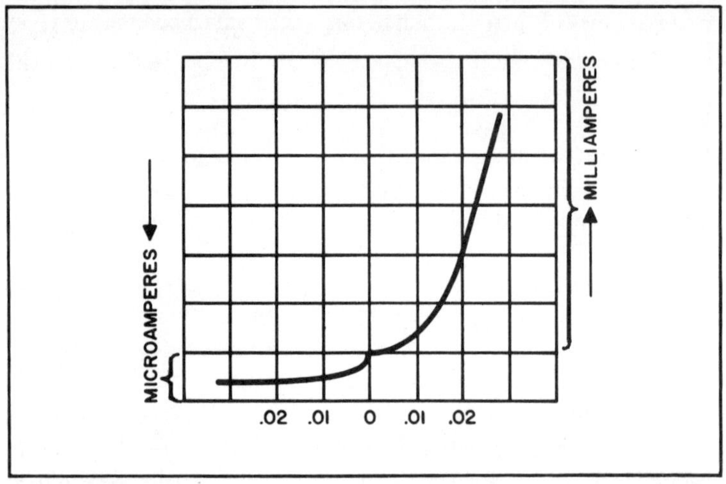

Fig. 2-15. Forward and reverse currents of a semiconductor diode. As the forward bias is increased, the current increases gradually (at first) but a point is soon reached where the current takes a sharp rise. The reverse current is much smaller than the forward current, is measured in microamperes. In this diode the forward current is in milliamperes. The difference is much more marked than would seem offhand since it takes a thousand microamperes to make up one milliampere.

when the battery is transposed. As the voltage is increased, we get a very small rise in reverse current. Note, in the instance of reverse bias, that the reverse current remains fairly uniform for a large part of the curve.

It is also very important for you to see that the forward current is plotted in milliamperes and the reverse current in microamperes. Technically, both curves should have been plotted in the same units but even here you can see that relatively, the forward current is far more substantial than the reverse current.

While the graph of Fig. 2-14 is useful as a comparison between forward and reverse currents, it does not supply an adequate picture in the region near zero voltage. It is close to this point that the potential barrier exists and, graphically, we are interested in seeing what happens to it. Figure 2-15 is an enlargement of the graph close to the zero point. As we increase our forward voltage, from zero to .01, there is a bare increase in current. This tells us that we have not yet overcome the potential barrier. But look what happens to the curve as we reach .01 volt. The current rises dramatically. Since this occurs at approximately the .01-volt point, we can assume that this was the value of the potential barrier.

Now let's examine the reverse current in the same way—near

the zero-volt point. Here there is a small, but steady, increase in reverse current, as we go from zero to .01 volt. But beyond this point the current levels off.

HOW HIGH IS HIGH?

With the pn diode, as with everything else, a limit exists somewhere. Figure 2-16 illustrates this very nicely. As we increase the reverse bias, the reverse current remains quite uniform.. But when we hit a certain voltage point we get an extremely abrupt increase in current. Increasing the voltage beyond this point has no effect, since the maximum reverse current is now going through the diode.

To understand what has happened we must go back to our potential barrier. With the application of reverse bias, the potential barrier increases. But this barrier is a voltage, and so there is an electrostatic stress between the two ionized regions—one in the p-type material and the other in the n-type. This stress—or

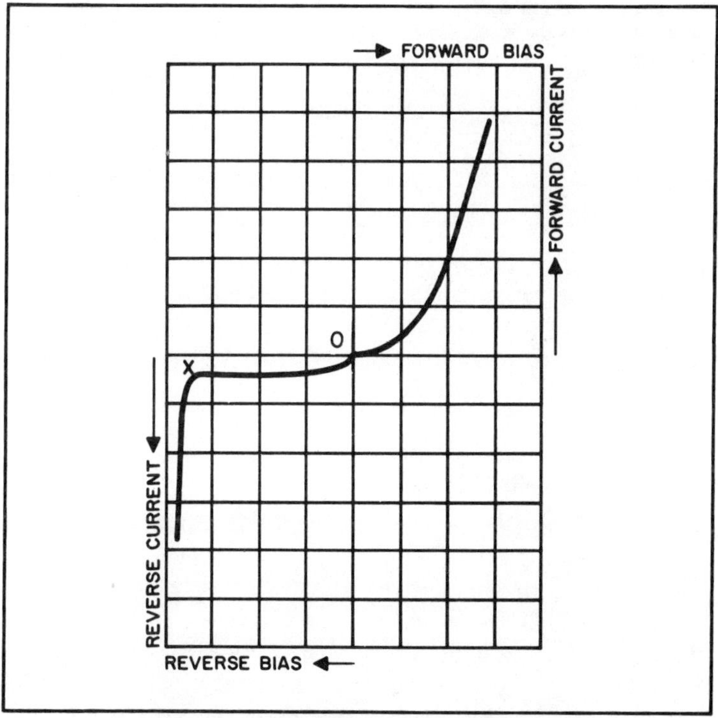

Fig. 2-16. If the reverse bias is increased, a value is reached at which the reverse current increases drastically. This is at point X in the graph.

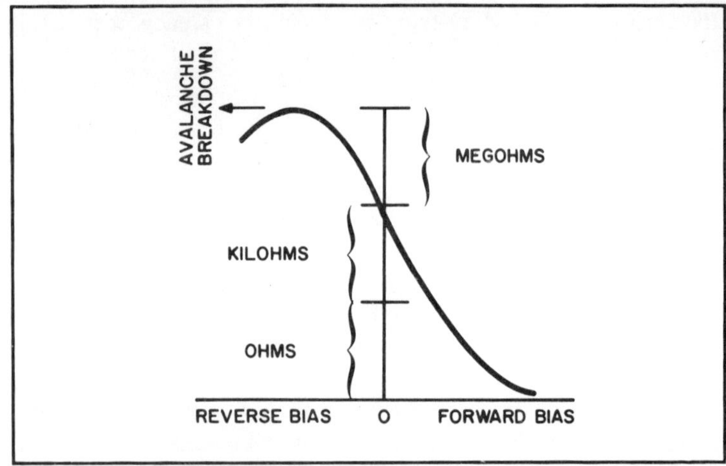

Fig. 2-17. Diode characteristics can be interpreted in terms of resistance. An increase in forward bias results in a decrease in resistance. An increase in reverse bias raises the resistance.

field—is a strong force. Electrons that would normally be quite content to remain in the outer orbit of their atoms, are pulled away. Of course, in moving out of position, they leave holes thus creating quite an opportunity for other electrons to get moving. Still more electrons are added to the general stampede through collision—electrons being knocked out of orbit, just as a pedestrian could get bowled over by a mad rush of people.

What we get now is a flood, an avalanche of electrons—and the graph in Fig. 2-16 shows this very well. The whole chain of events we have described is called avalanche multiplication. The critical point at which this multiplication of electrons in motion starts, producing a current that is independent of voltage, is called avalanche breakdown. Point X on the graph is the avalanche breakdown point.

RESISTANCE CHARACTERISTIC

We've been talking about voltage and current, but whenever we do, you can be sure that you will find resistance lurking somewhere about. Figure 2-17 is the graph of the resistance characteristic of a semiconductor diode. But while we refer to it as a resistance characteristic, it is really a measure of the effectiveness of the potential barrier. Examine the curve in the region of zero volts. Put your finger at the zero mark and then move up on the vertical line until you reach the curve. The value here is somewhere between kilohms and megohms. From here we can move down on the curve in the direction of forward bias. As the forward bias overcomes the potential barrier, we get a decreased opposition to current flow. This is just as though we had reduced the resistance at the junction.

Moving in the other direction, you can see that the resistance curve rises to a peak. This peak is really a plateau, for the resistance remains high, remains uniform. But as we move further to the left we are moving in the direction of higher reverse bias. At some voltage we will reach the breakdown point, and, of course, the resistance will take a precipitous drop.

A FEW IMPLICATIONS

A graph is more than just a pretty picture and there is a lot more significance to the curve in Fig. 2-17 than its resemblance to a scenic railway. What we are really saying is that we can control *resistance* by means of voltage. Now this might not seem like a startling innovation until we remember that we ordinarily think of *current* as being controlled by voltage. Ohm's law emphasizes this for us in no uncertain terms. But in Fig. 2-17 we have managed to adjust the resistance of our diode in two ways. The first depends on polarity—that is, whether we are forward- or reverse-biasing our diode. And the second is predicated on the amount of bias voltage used.

In the meantime, you may have noted that, while we discussed the effects of an excessive reverse voltage, no mention has been made of excessive forward voltage. This bit of information is being reserved for the discussion on transistors.

DIODES HAVE MANY USES

The most obvious application of the semiconductor diode is to change an alternating current (ac) into pulsating direct current (dc). As long as we don't put too much reverse voltage across the diode, causing an electron avalanche, the diode will pass current effectively in only one direction. Yes, a tiny current will pass in the reverse direction, but it is negligible; and no, the diode does not quite perfectly conduct in the forward direction, but it's almost perfect.

There are other applications for the semiconductor diode, though, besides the conversion of ac to dc. It's truly amazing what these seemingly simple devices can be made to do. Let's look briefly at the most common uses.

Rectification

Figure 2-18 shows what happens when we put an alternating current through a diode. Half of the cycle is effectively cut off. Alternating current flows in two directions, back and forth—first so the diode is forward biased, and then so it is reverse biased. Of course, current can't go in the reverse direction, try as it might. So half the cycle just disappears! What we have left is pulsating dc, which changes in intensity but always goes the same direction in the circuit. The pulsations can be smoothed out using filters (which we will not discuss here), leaving ordinary dc just like the kind we get from a storage battery.

Before semiconductors were invented, tubes were used to rectify ac. Besides being ungainly, tubes were hot! You would have to build a special power supply just to make the tube do its job. Later, mercury-vapor tubes were used which did not need the separate filament power supply. Tubes had a large voltage drop in the forward direction, and so they worked well only with dangerously high voltages. Semiconductors are small, much more effi-

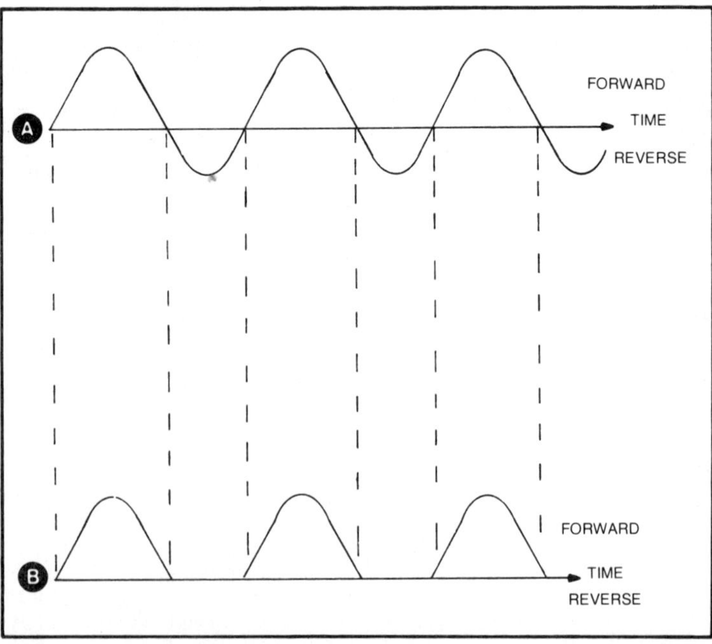

Fig. 2-18. When alternating current is passed through a diode, it can only go in one direction. This results in pulsating direct current. This effect, called rectification, occurs with all diodes at low frequencies, and is one of the principal uses for the pn junction.

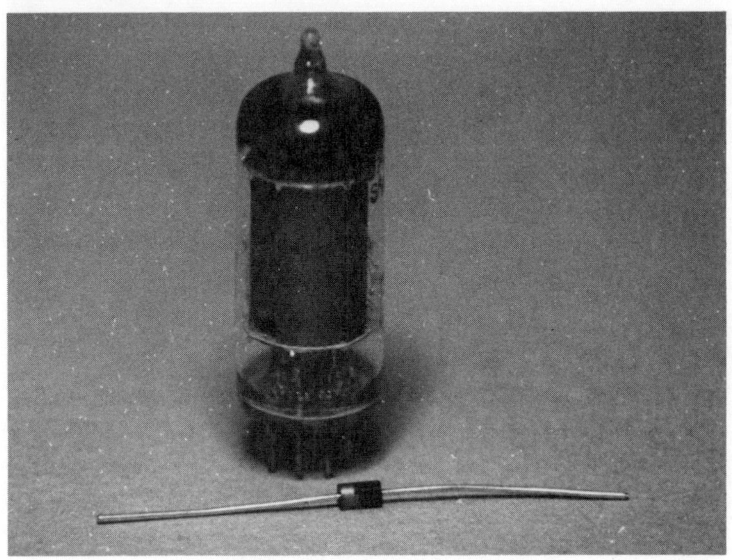

Fig. 2-19. Diodes are much smaller, and much more efficient, than their predecessor, the rectifier tube!

cient, and will work at low or high voltages. If you have a very old piece of radio equipment or an old television set, open it up and look for the rectifier tubes, and you can see the difference for yourself. Or, take a look at Fig. 2-19.

Detection

In a radio receiver, detection refers to the separation of the voice from the carrier wave of an amplitude-modulated (AM) signal. Does that sound complicated? Actually, detection is exactly the same thing as rectification, except that the frequency is much higher (kilohertz or megahertz rather than 60 hertz).

A voice-carrying AM signal looks something like Fig. 2-20A if we observe it on an oscilloscope. After passing through a diode, half of the signal gets chopped off (Fig. 2-20B). The carrier wave is shown as extremely rapid pulses, and the more gradual changes are voice frequencies. By using a capacitor, we can filter out the jack-rabbit carrier and leave the audio frequencies intact (Fig. 2-20C). Then we can amplify the audio using, well, transistors! We'll see how that's done later.

Light-Emitting Diodes

If you have an FM stereo receiver, maybe it has a little red light

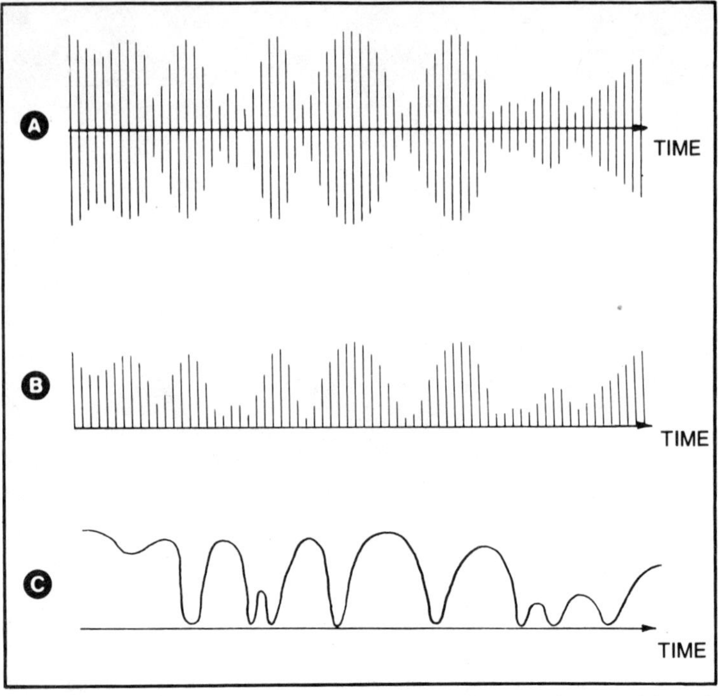

Fig. 2-20. Detection occurs when a high-frequency AM signal is passed through a diode. At A, typical AM signal as seen on an oscilloscope; at B, after passing through diode; at C, after signal is filtered, leaving only the audio information.

that tells you when you've tuned in a stereo signal. Or maybe you have a digital clock radio with red or yellow glowing numerals. Perhaps your watch has a luminous digital time display, but you have to press a button to read it. All these glowing things are—you guessed it—diodes! When current flows through a certain type of diode, it emits visible light. There are green, red, and yellow ones, and combination devices that can change color. Light-emitting diodes, or LEDs, are used in many electronic devices today.

LEDs last a long time and consume very little power. They are a vast improvement over the old incandescent lamp. There are two different ways of manufacturing LEDs (Fig. 2-21), known as point-source and area-source. The little red light in the FM stereo receiver is a point-source LED, which makes use of a lens to diffuse the light. Most digital displays employ the area-source LED. Seven such diodes, arranged to look like a square numeral 8, may be switched on and off to form numerals and certain letters of the alphabet.

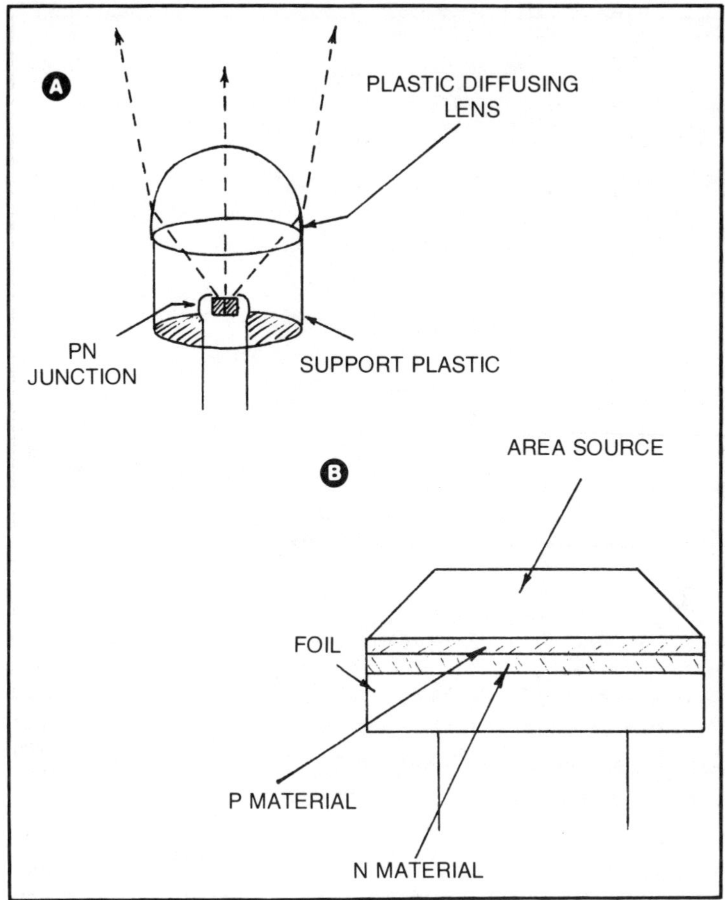

Fig. 2-21. At A, a point-source light-emitting diode. At B, an area-source LED.

Solar-Electric Diodes

What the diode can do, it can also undo! A good example of this is evident when we compare the LED to the solar-electric diode or solar cell. The solar-electric diode converts visible light into dc electricity. Figure 2-22 shows how this kind of cell is made. Using large batteries of such diodes, a small radio receiver or transmitter can be operated directly from sunlight. No energy shortage there! Conventional storage batteries are sometimes charged during daylight hours using solar-electric diodes. Then, their power is available at night. Solar-electric diodes react very quickly to changes in light intensity, and so they may be used to receive modulated-light signals.

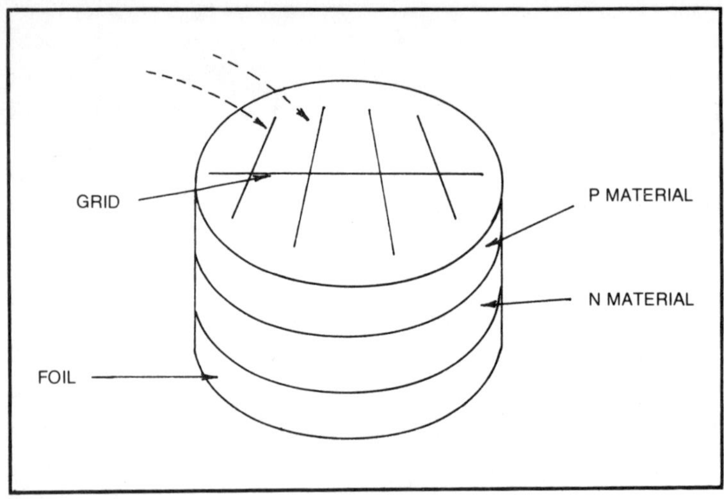

Fig. 2-22. Pictorial diagram of a solar-electric diode. Impinging light (dotted lines) creates a dc potential between the p and n materials.

Zener Diodes

The avalanche breakdown of a diode can be a useful effect, if it occurs when we want. The reverse voltage that causes this breakdown is very critical, and always the same for a given diode. Some kinds of diodes have much greater avalanche breakdown points than others. Zener diodes are deliberately manufactured to have a certain breakdown voltage, which varies in practice from a few volts to around 200 volts.

When we connect a Zener diode in the reverse (ordinarily non-conducting) direction across a source of voltage, as shown in Fig. 2-23, the voltage between the output terminals can never exceed the avalanche breakdown rating of the diode. Of course, this might be academic, if the Zener voltage is much larger than the input voltage. But by using a series resistor and an input voltage greater than the Zener voltage, we can achieve very precise voltage regulation at the power supply output. We might design a power supply to deliver, say, 17 volts dc, and place a 12-volt Zener diode across the output terminals. The resulting output voltage will remain at 12 volts even when the load is changed. Another benefit of using a Zener diode in the output of a power supply is the elimination of transients, or high-voltage surges caused by lightning and other power-line irregularities.

The power ratings for Zener diodes vary between approximately a quarter watt and 50 watts. Zener diodes have taken the place of regulator tubes in modern power supplies. Regulator tubes were very inefficient and somewhat finicky devices; Zener diodes do their job quietly and cooly. Zener diodes function very well for purposes of biasing high-power tube amplifiers (but we aren't concerned with tubes in this book!). When several different voltages are wanted from a single power supply, Zener diodes may be connected in series across the supply output and the various voltages will appear across the diodes with properly chosen ratings.

GUNN DIODES

In certain situations, diodes will oscillate at high frequencies when the right amount of forward current flows through them. This is called the Gunn effect, after its discoverer, J. B. Gunn of IBM. You might suppose that this effect would be a nuisance, and of no practical value. But specially made diodes are used as signal generators at microwave frequencies. Gunn diodes are not very efficient; only a tiny percentage of the power they consume actually results in microwave oscillation. Still, as much as 1 watt may be obtained at frequencies of 10 GHz by means of a Gunn-diode oscillator. This is an entirely practical amount of power in this frequency range.

Specialized diodes called IMPATT (impact-avalanche transit time) devices are used as amplifiers at microwave frequencies. Another kind of diode, called a tunnel diode because of its strange voltage/current characteristic curve, is also sometimes used as an oscillator and amplifier for microwave signals.

Varactor Diodes

We are now beginning to see some of the truth in the statement, "One experimentalist will keep a dozen theorists busy!" No one would have guessed, just by cogitating while sitting in a chair,

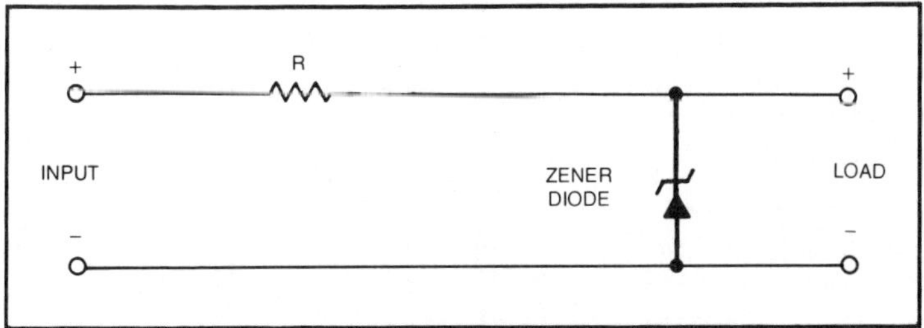

Fig. 2-23. A Zener diode may be connected across a power supply, in conjection with a dropping resistor R, to provide precise voltage regulation.

that diodes might ever be used as oscillators. But experiments showed, quite accidentally, that they could. Similar experiments have also shown that diodes make excellent variable capacitors. Nowadays, FM radio transmitters make use of this effect of the diode in the varactor, short for "variable reactor" since the capacitive reactance varies with the bias.

Such diodes are not employed as rectifiers. Instead, after connecting one to an oscillator, the frequency of oscillation may be changed simply by applying a variable voltage to the varactor. The voltage is applied in the reverse direction. As the voltage is increased, the capacitance between the p and n layers gets smaller.

Varactor diodes react very fast, and by using them in oscillators, excellent frequency modulation is attainable. An FM oscillator using a varactor diode is called a voltage-controlled oscillator for this reason.

PIN Diodes

We hope you haven't gotten the idea that diodes may be used only as rectifiers, detectors, energy converters, oscillators, amplifiers, voltage regulators, and capacitors. Their scope is not that limited! Specially made diodes are also used as high-speed switches, similar to relays but with no moving parts to wear out. These diodes are called PIN diodes, which stands for the materials from which they are made. An intrinsically pure (i) silicon layer is

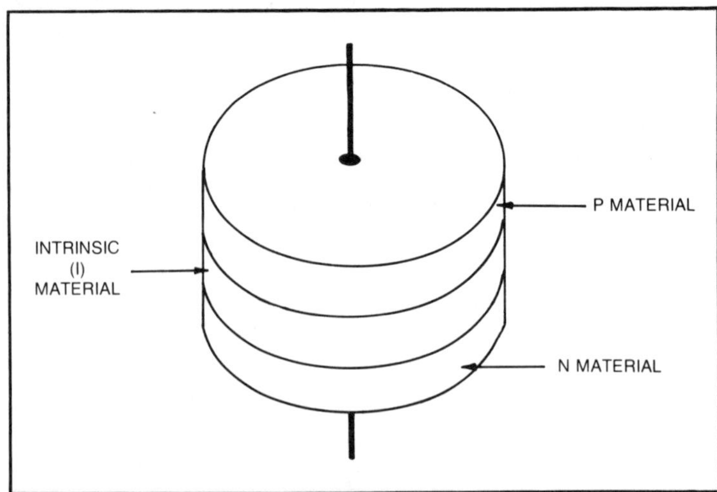

Fig. 2-24. The PIN diode has a layer of intrinsically pure (i) silicon between the p and n layers to reduce the capacitance at high frequencies.

sandwiched between the p and n layers (Fig. 2-24). This reduces the effective capacitance between the two semiconductor layers, in turn minimizing the amount of high frequency leakthrough.

When a PIN diode is forward biased by a source of dc, it will conduct. Then it is reverse biased, or when there is no voltage applied to the diode, it acts as an open circuit with excellent isolation properties. At high frequencies, an ordinary diode will pass significant energy even when it is not conducting dc. We find PIN diodes in antenna switching circuits of transceivers, where a single antenna must be transferred from a transmitter to a receiver and back again many times, with high speed and reliability. PIN diodes have taken the place of the bulkier and less reliable relay for this purpose.

PIN diodes may also be used for building attenuators, devices for reducing the strength of incoming signals by a precise amount over a wide range of frequencies.

Frequency Mixers and Multipliers

The list of applications for the seemingly humble diode goes further! Certain pn-junction devices are used in solid-state circuits to mix two signals together, obtaining the sum and difference frequencies. For some time, most receivers have used the superheterodyne principle. The incoming signal is mixed with another variable-frequency signal to produce an output with constant frequency, which is much easier to process with a high amount of gain. Such mixer circuits originally used tubes, and later transistors. But recently, diodes have been found to do this job with extreme simplicity and efficiency. Hot-carrier diodes (HCDs) are designed especially for this application.

The HCD has lower resistance and higher breakdown voltage than an ordinary diode. It allows less current to flow in the reverse direction. The HCD is a sort of super diode, close to perfection.

Another strange kind of diode, called a step recovery or snap diode, is used for the purpose of frequency multiplication. Using these diodes, extremely high-frequency signals may be generated with quite ordinary transmitting equipment. For example, the harmonics of a two-way VHF transceiver may be used to generate a signal in the 10-GHz band for amateur use by means of snap diodes in the output circuit.

THE TRANSISTOR COMETH

Since the very first page of this book we have implied a promise to talk about transistors, a promise that it would seem has not yet been fulfilled. We have spent quite a lot of time just talking about diodes. Transistors are, however (as you shall soon see) just one stage beyond the diode in the evolutionary process of semiconductors. We need to know about diodes before we can really talk in a meaningful way about transistors.

QUESTIONS

1. What is reverse bias? Forward bias?

2. What is a negative ion? A positive ion?

3. Describe ionization at the pn junction.

4. What is the potential barrier? How can we increase it? How can we decrease it?

5. What is electron injection?

6. What is avalanche voltage? Avalanche current? What conditions produce avalanche current? How can we take advantage of it?

7. What is the name of the battery connected to the pn diode?

8. How does forward current differ from reverse current in terms of magnitude? What types of diode are deliberately reverse-biased?

9. What is the relationship between the resistance of a pn junction and the potential barrier?

10. What type of field is set up between blocks of p- and n-type germanium in close contact?

Chapter 3

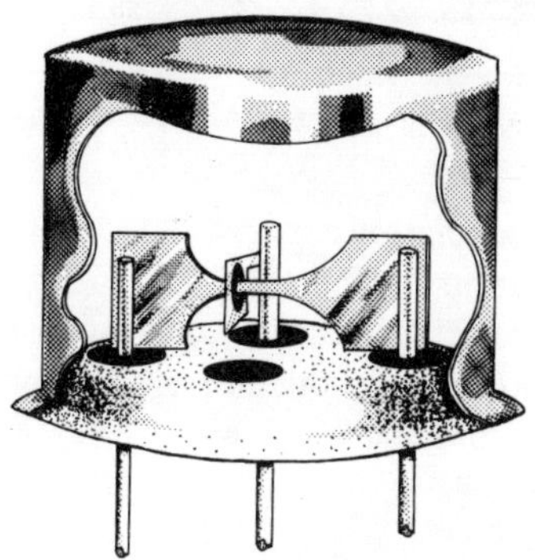

Enter the Triode

WE ARE NOW IN THE HAPPY POSITION OF A CONTRACTOR WHO-arrives on the scene and finds all his building materials on hand. We can start the labors of this chapter with the two diodes shown in Fig. 3-1. Here we have a pair of pn diodes (we could call them np diodes just as correctly) and we have placed them so that the two n-sections are in close contact.

Our two n-sections, though, are fairly identical, so there is no reason why we cannot have one single block of n-type germanium or silicon instead of separate pieces. Figure 3-2 shows our new arrangement. Other than the unification of the n-section (and that was no radical move) we have made no changes whatsoever.

What do we have now? Just three blocks of material—somewhat giving rise to the idea of a sandwich, but rather more difficult to manufacture. However, two of our blocks are still identical, so to keep from compounding confusion, let's give them names. The one at the left is called the emitter, the one at the right the collector, and separating the two is the base. This is a transistor, and just as we informed you earlier, it can be regarded as a pair of diodes placed back to back.

BIASING THE TRANSISTOR

As long as diodes are separate entities they have a sort of independence. Consider the transistor in Fig. 3-2, though. How do we bias it? We can connect a battery between the base and the

Fig. 3-1. Our first step toward the transistor is to use a pair of pn diodes back to back.

emitter. We can immediately recognize this (in Fig. 3-3) as forward biasing, since the plus terminal of the battery connects to the emitter (p-type material) and the minus terminal to the base (n-type material). Disregarding the collector for the moment, we have a single forward-biased diode.

CHANGING THE BIAS

Our forward bias battery is connected between the base and the emitter. Changing the bias is no great problem, since we can follow exactly the same techniques described earlier in Chapter 2. The idea is repeated in Fig. 3-4. Adjusting the moving arm of the potentiometer gives us any value of bias we want, within the voltage limits of our battery. In the drawing of Fig. 3-4 we are assuming a battery voltage of 2.

We have another technique for varying the bias, although it may not be so readily apparent. The technique is illustrated in Fig. 3-5. Here we have an ac voltage source in series with the battery. What is this voltage? It could be supplied by a transformer, a signal generator, or it could be a broadcast signal. But regardless of how we obtain it, suppose we assume that the voltage has a peak-to-peak value of 1. This means that the maximum positive peak is one-half volt. Similarly for the maximum negative peak, as shown in Fig. 3-6.

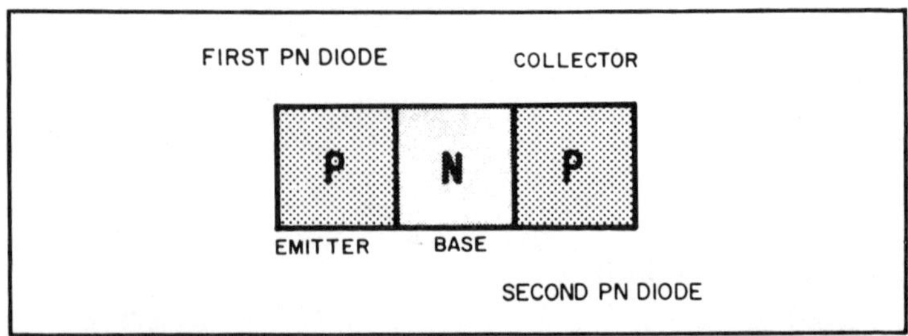

Fig. 3-2. We still have a pair of pn diodes but we have now joined the two n-sections into a single unit.

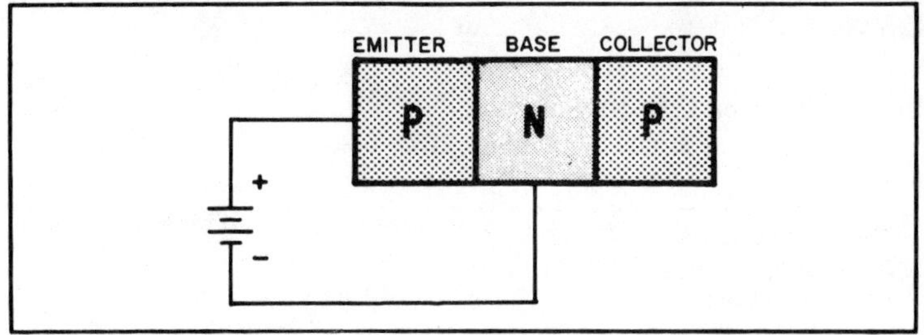

Fig. 3-3. With the positive terminal of the battery connected to p-type semiconductor and the negative terminal to n-type, we have forward-biased the diode.

What effect will this have? It is just as though we kept varying the potentiometer arm in Fig. 3-4 at a frequency rate equal to that of the input signal. When the signal reaches its positive peak, the two voltages ac and dc, will add, and the total bias will be the two volts contributed by the battery and the half volt supplied by the ac source. The two voltages, ac and dc, at this time, are series aiding. As the ac voltage climbs down from its maximum positive peak and moves toward the zero line, we get a reduction from the maximum bias of 2.5 volts. When the ac voltage is zero it is as though it has been effectively removed from the circuit, and the only bias is that contributed by the battery—2 volts. But once the ac voltage crosses its zero line, it moves in a negative direction, and now, instead of aiding the battery, it opposes it.

At its maximum negative peak, we have one-half volt ac. This voltage, whose polarity at the moment is contrary to that of the bias battery, has the effect of subtracting from it. And so our bias now drops to 1.5 volts. The total value of forward bias swings from 1.5 to

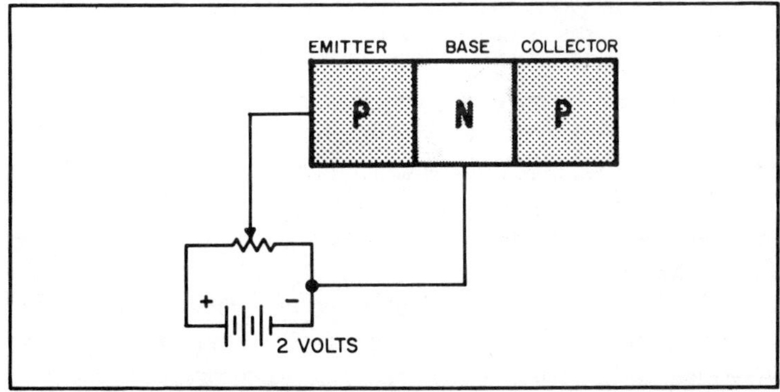

Fig. 3-4. In this case, the forward biasing can be varied from zero to a maximum of 2 volts.

2.5 volts. At all times, though, our pn-section is forward-biased, to a greater or less degree, depending, at any instant, on the algebraic sum of battery and ac voltage.

What effect will this changing bias have? The barrier potential of our pn-junction—the junction between the sections we now call the emitter and the base—will vary. The greater the bias, the lower the barrier potential, and the greater the amount of current flow.

SIGNAL VOLTAGE AND BATTERY VOLTAGE

There is a very interesting point to note here. Compare the values of the signal and bias voltages. At no time is the signal voltage greater than the voltage supplied by the battery. This means that current flows at all times in what we can now regard as our base—emitter circuit. Our constantly forward-biased base—emitter circuit consists of n-type material (the base), p-type material (the emitter), and all components and voltages connected in series between these two.

IMPEDANCE

If we were to have a purely resistive circuit, such as a battery shunted by a resistor, the opposition to the current flow would be simple and well-defined. It would consist of the value of resistance in the circuit (assuming that wire resistance, resistance of soldering joints, resistance of the battery, etc., would be negligible). But, add some coils or capacitors, and change that dc to ac, or put a generator (or other ac voltage source) in series with the battery, and the total opposition to current flow becomes a little more complicated. The frequency of the ac and the coils and capacitors in the circuit, all contribute to an opposition effect we call reactance. Combined vectorially with resistance we get the total opposition to the flow of current. We call it impedance. Like the reactance and resistance of which it is formed, impedance is measured in ohms.

When impedance is specified in actual ohms, there is no doubt as to what is meant (if the frequency is mentioned along with the impedance). But quite often terms such as low impedance and high impedance are used. These words are meaningless unless we have some reference.

This might seem strange until we compare dc and ac. For resistance we specify the value in ohms since we know it isn't going to change, whether we use dc or ac. But impedance, also in ohms, doesn't have such fixed habits. Its value fluctuates with frequency, while for dc, it has no meaning at all.

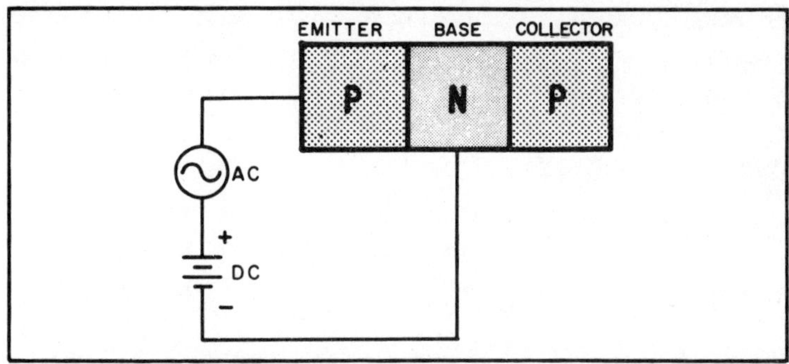

Fig. 3-5. Instead of using a potentiometer to change the amount of forward biasing, we can use a signal voltage, represented here by the symbol for an ac generator.

IMPEDANCE REFERENCE

Consider the vacuum-tube output circuit of an ordinary radio receiver. The plate is referred to as a high impedance point. But the control grid has an even higher impedance. And the voice coil circuit is called low impedance. In these three circuits—grid, plate, and voice coil, what determines impedance? It is the amount of *relative* current flow. Thus, the control grid circuit has little or no current flow. But small current means high opposition or impedance. The plate circuit has much more current flowing in it (compared to the control grid) hence has a lower impedance. And the voice coil circuit, thanks to the assistance of the stepdown transformer, has more current flowing in it than either the plate or control grid, hence has the lowest impedance of the three. We always associate low impedance with current flow—high impedance with very small or no current flow.

Getting back to our forward-bias base—emitter circuit, we can temporarily call it a low-impedance circuit—temporarily, because we have established no other operating circuit as yet to which we can compare it. Also, in Fig. 3-5, we have not as yet inserted

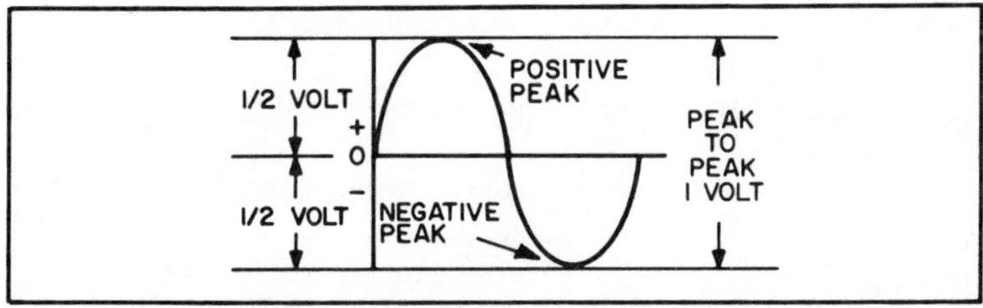

Fig. 3-6. Both the positive and negative halves of an ac input have an effect on the forward biasing.

reactive components (such as coils or capacitors) in our base-emitter circuit, hence practically all of the opposition to the forward-biased current is resistive.

AND STILL MORE BIAS

Up to now we have been working with just one part of our transistor—the emitter-base circuit. We can add another battery, as shown in Fig. 3-7. This battery is connected between the collector and the base. But just how did we get our collector and base? The base is n-type material and the collector is p-type. This, as we learned earlier, is one of the diodes we used to make our transistor. But note how this part of the transistor is biased. The negative terminal of the battery is connected to the collector (p-type material) while the positive terminal goes to the base (n-type). What we have here is reverse biasing. And, just as in the case of the base-emitter circuit, we can vary the amount of reverse biasing by using a potentiometer, or by putting an ac voltage in series with the reverse-biasing battery.

INPUT AND OUTPUT

With the help of Fig. 3-8 we can take another forward step. The base-emitter circuit now has another name—input. And the base-collector circuit can also be identified in another way—output. Figure 3-8 is interesting in that it emphasizes the fact that the base is common to both input and output circuits. But, unlike the elements in a tube, separated by space, our three transistor elements, emitter, base and collector, are in very close contact.

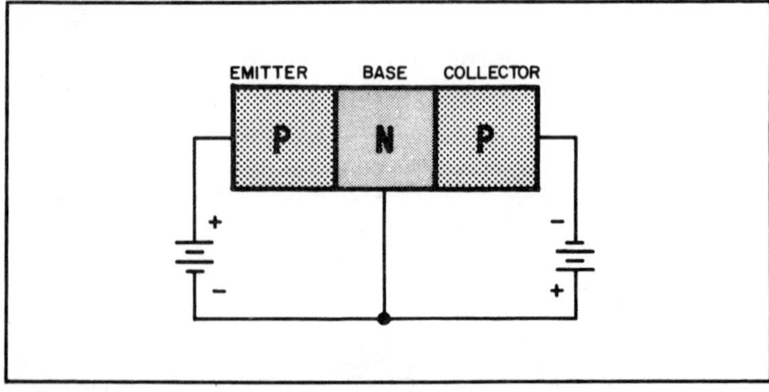

Fig. 3-7. Our forward-biasing technique remains the same, but note that we have added another battery and that this battery forms a reverse-biased connection between base and collector.

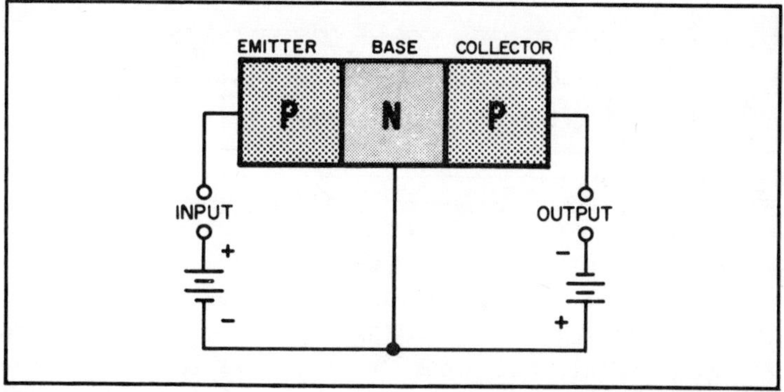

Fig. 3-8. We can break open the circuit of Fig. 3-7 for input and output, as shown here.

THE TRANSISTOR SYMBOL

Just as a child in early school grades gradually puts away his blocks and takes to the use of more convenient numbers and letters, so too can we dispense with our block-diagram arrangement of the pnp transistor and use its electronic symbol. This is shown in Fig. 3-9. The base is marked with the letter B, the emitter with E, and the collector with C. The emitter and collector are always drawn at angles to the base. The emitter always has an arrowhead. For a pnp transistor, the first letter (p) always represents the emitter, the second letter (n) is the base and the third letter (p) is the collector.

THE NPN TRANSISTOR

In constructing our transistor we started with a pair of pn diodes. We joined the two n-sections and so the terminal point of this activity was a pnp transistor. But Fig. 3-10 shows that we selected just one of two possible ways of joining the diodes. We could have butted the two p-sections, and thus, as the illustration indicates, could have produced an npn unit.

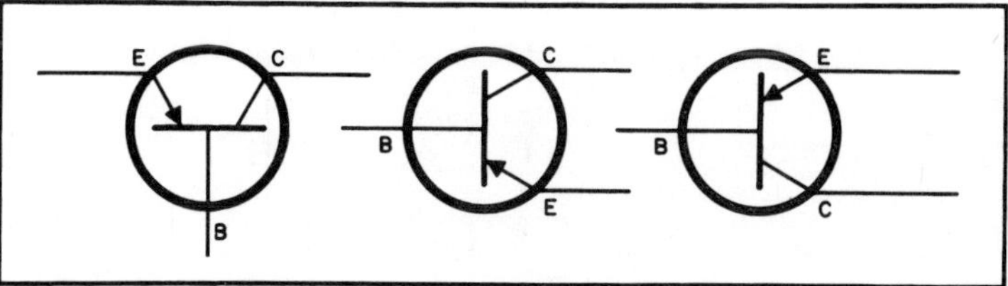

Fig. 3-9. Here is the symbol for the pnp transistor. The symbol can be put in any position and we can transpose the emitter and collector leads for our convenience in circuit diagrams.

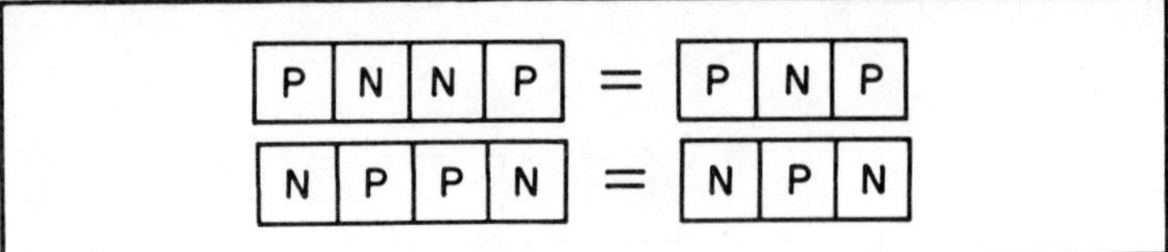

Fig. 3-10. Formation of the pnp and npn transistors follows the same techniques in both cases.

Biasing for the npn transistor follows the same sort of arrangement we used for the pnp—that is, the input is forward-biased (as shown in Fig. 3-11) while the output is reverse-biased. Be careful, though. To make the npn transistor, we transposed our pair of diodes, so now we must reverse the polarity of our bias batteries. Other than that, there is no difference between the two types of transistors.

ELECTRONIC SYMBOLS

We have the electronic symbol for the npn transistor in Fig. 3-12. We have also repeated the symbol for the pnp so that you can compare the two. The only difference is in the arrowhead. In the pnp unit it points inward, while in the npn it points outward.

Various letters are often used to represent transistors. The letter T (for transistor) is occasionally used. Gradually, however, it has become popular to use the letter Q, since it stands for nothing else! We will use Q throughout this book.

ADDING COMPONENTS

We now have our transistor, but, to add a few final touches to the picture, suppose we add a pair of resistors (as in Fig. 3-13). And

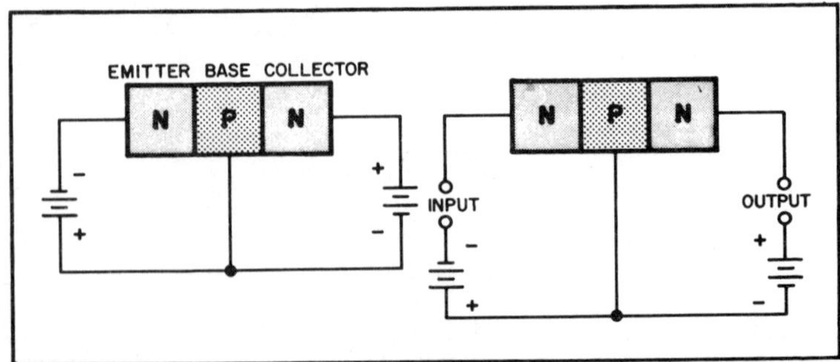

Fig. 3-11. The formation of the npn circuit follows the same approach we used for the pnp in Fig. 3-8.

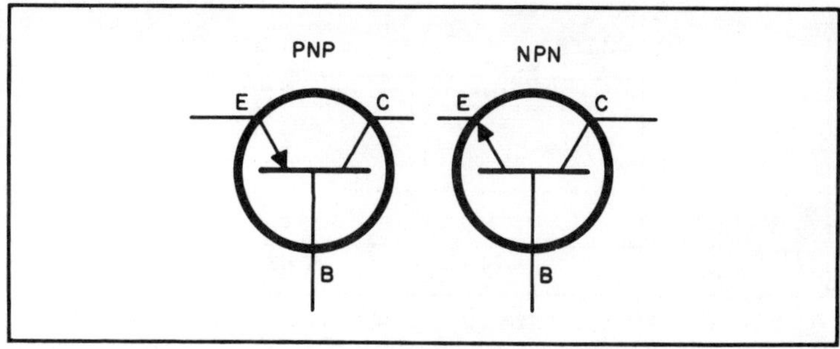

Fig. 3-12. Symbols for the pnp and npn transistors. They are almost identical. The only difference is in the emitter element. For the pnp the arrow points inward to the transistor; for the npn it points outward.

so, with these few components, for both types of transistors (npn and pnp) we are ready to learn how the transistor works and, having learned that, find out how we can put the transistor in key spots in a variety of jobs.

HOW THE TRANSISTOR WORKS

Up to now our block diagrams of the transistor created the impression that each of the blocks—emitter, base and collector—contained equal volumes of material. To have our two diodes work together as a transistor, let's modify our diodes and our thinking at the same time. We now want you to think of the base as an extremely thin section, compared to the emitter and the collector.

Continuing with our analysis of the transistor as an arrangement of diodes, look at Fig. 3-14. What we have here is a pnp transistor but we are using diode symbols. If we disconnect the lead going to the base, you can see quite readily that the two diodes and their biasing batteries form a series circuit. And since we have a series circuit, the current will be the same in all parts of the circuit.

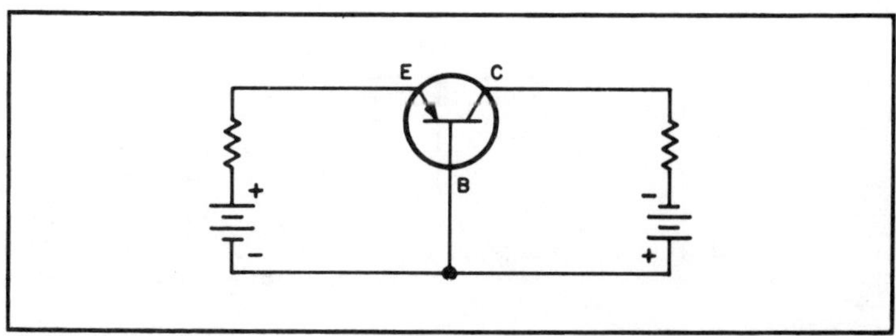

Fig. 3-13. With the addition of a few resistors, our elementary transistor circuit begins to take shape.

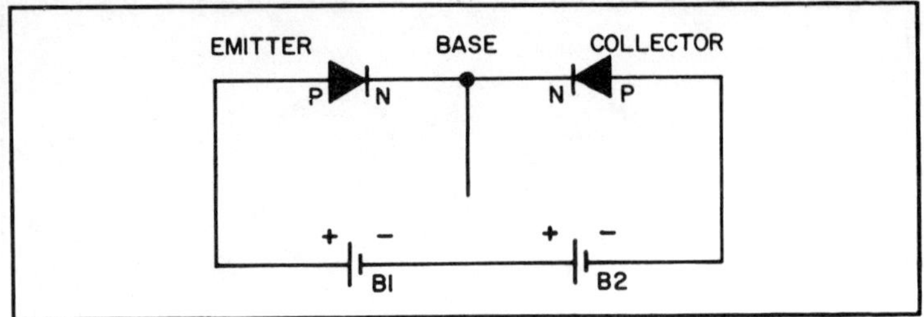

Fig. 3-14. The diodes and their biasing batteries form a series circuit.

Also, take a good look at the batteries. They are connected in series aiding.

Just so that we won't be accused of showing a preference for the pnp type, we also have an npn unit in Fig. 3-15. It is exactly the same as Fig. 3-14, except, of course, the diodes have been turned around, and so have the batteries. But the two diodes are still in series and the batteries still form a series-aiding pair.

What about current flow in the circuit of Fig. 3-15? Assuming current flow to mean electron flow, we can start at the negative terminal of B1. The current will move through the block of material marked n (the emitter), through the two combined (but very thin) p-regions (the base), into the n-region (the collector), and then from that region through B2 and B1 back to the starting point again.

Although we started this explanation by assuming that current first moved out of the negative terminal of B1, remember that this was just a matter of convenience. The current in a series circuit is a bit like a merry-go-round. Which part starts first?

UNFAIR! UNFAIR!

The circuits of Figs. 3-14 and 3-15 aren't really playing the

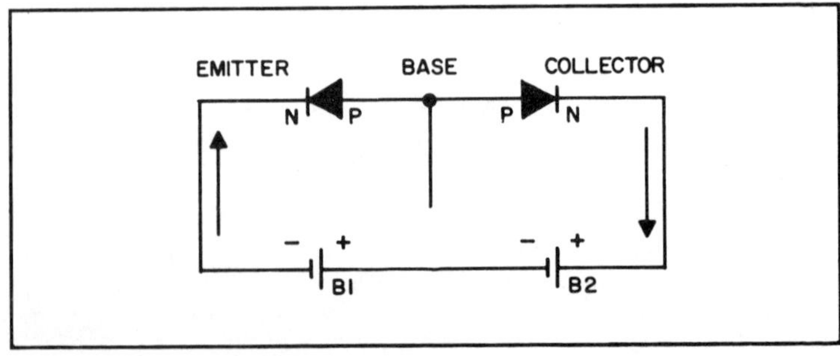

Fig. 3-15. This circuit is similar to that of Fig. 3-14 except that both batteries and diodes have been transposed.

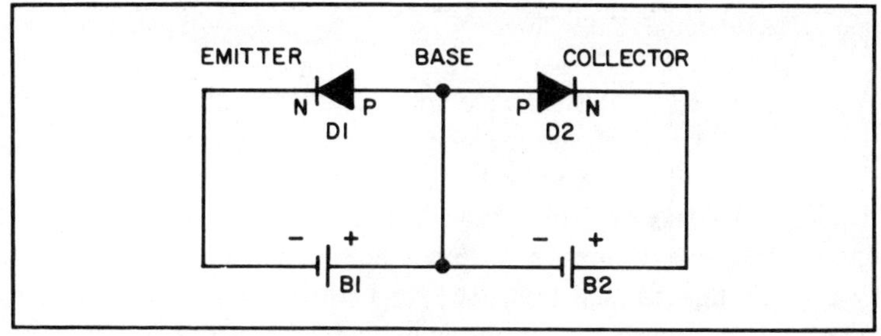

Fig. 3-16. With the connection of the base lead we provide a current path between the base and battery B1.

game and we're violating some of our own rules. We've talked at great length about the diodes being forward- and reverse-biased, but the disconnected base lead in Fig. 3-15 isn't cricket. So, as shown in Fig. 3-16, let's connect the base lead to the junction of the two batteries.

Now you must admit that everything is quite proper. We have our two diodes—one forward-biased, the other reverse-biased. What effect will this have on the current?

Offhand, you might argue that the two diodes could be represented by two resistors. Diode D1 in Fig. 3-16 is forward-biased, the current should be comparatively large, hence we could say it is equivalent to a small resistance. Diode D2 is reverse-biased, the current should be much smaller than through diode D1, hence its resistance is much higher. But D1 and D2 are in series. If we have a small resistor (say of 1 ohm) in series with one having a large value (say, 1,000 ohms) we can practically forget about the 1-ohm resistor as far as the total series resistance is concerned. But D1 and D2 are not resistors. They're a pair of diode elements constructed in a very special way.

Since we cannot consider D1 as a low-value resistor only and D2 as a high-value resistor only, let's see just how they do differ. Let's analyze them from the viewpoint of an electron current. In Fig. 3-16, D1 is forward-biased. To trace the path of electron current flow, suppose we start at the negative terminal of B1, since that is just as convenient a starting point as any. The electrons move along to the first diode, D1. Because this diode is forward-biased, its potential barrier is pleasantly low, and so the electrons move easily enough into the base.

Our electrons are now faced with the necessity for making a decision. They can either travel down the wire to the positive terminal of B1 or they can make an effort to flow through D2 to the positive terminal of B2. Now everything we have told you up to now would make you think that there really is no choice. Going back to the plate of B1 we have a nice solid wire—an easy path for electrons. And, to clinch the argument in favor of such a path, D2 is reverse-biased and we know that it has a high potential barrier that opposes the passage of electrons. Yet, despite both of these factors, most of the electrons move through the second diode (D2) and from there through B2 and then back to battery B1. Since this is so contrary to the way the circuit would behave if we had a low-value resistor in place of D1 and a large resistor in place of D2, let's investigate a bit further.

We can do this by moving on to Fig. 3-17. We still have our two diodes, but now the base region is extremely thin. Furthermore, the base is common to both diodes. It is shared by D1 and D2. As before, when the electrons get into the base region, they have a choice of going down the solid conductor to the positive terminal of B1 or, seemingly, struggling through the collector n-region.

To understand what is happening, let's go back to Fig. 3-15 for a moment. B1 and B2 are in series. The total voltage of these two batteries (because of the series connection) now exists between the n-type collector and the n-type emitter. And, as we saw way back in Chapter 1, an electron likes nothing better than to move along the lines of force of an electric field. This is the same reason why electrons will move through a vacuum from a hot cathode (and sometimes a cold one) to a positive plate. As soon as the electrons find themselves in the base region, they feel the pull of the positive terminal of B1. But the voltage of B2 adds to the voltage of B1 to provide a strong electric field. So the potential barrier is overcome.

However, the base is still a dividing point for the electron current. Most of the current , though, proceeds through diode D2,

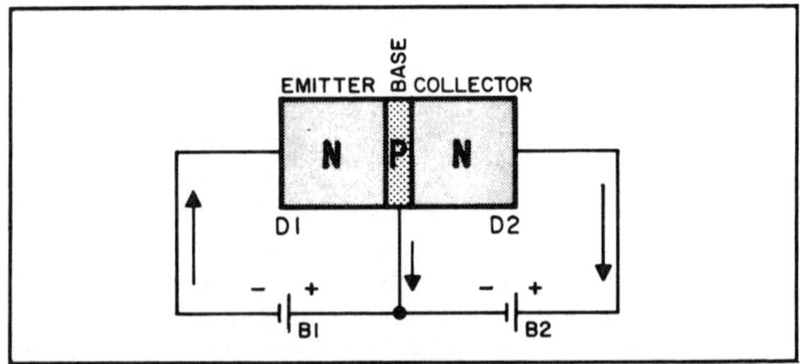

Fig. 3-17. Paths of current (electron) flow in the npn transistor circuit.

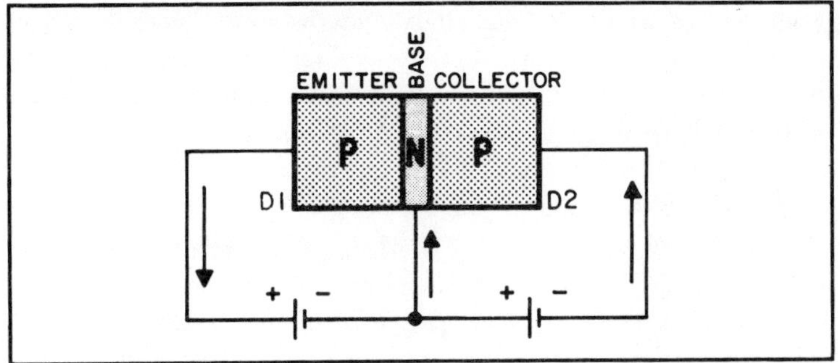

Fig. 3-18. Paths of current (electron) flow in the pnp transistor circuit.

with a much smaller amount flowing back to battery B1.

Our discussion of Fig. 3-17 centered around the npn transistor. What happens in the case of the pnp type? Exactly the same thing, except that the electron current flows in the opposite direction (Fig. 3-18 illustrates this). The arrows, representing the paths of electron flow, are opposite in Figs. 3-17 and 3-18.

NAMING NAMES

Not too long after the triode vacuum tube became popular, engineers found many reasons for giving the three elements added company. Grids were inserted, resulting in tetrodes, pentodes, and pentagrids. Probably the most widely-used receiving tube today is the pentode.

Like the tube, the transistor came into being as a three-element device. The transistor, though, is following its own evolution and since it is an entity in its own right, we have no reason for thinking that the transistor will follow the path trod earlier by the tube. At this time, the most widely used transistor is a triode, although other types are available. The triode (tri, the prefix, means three) is a three-element device.

We have a name for each element in our transistor—emitter, base and collector. The word emitter would seem to indicate that electrons are boiled out, much as in the case of a hot cathode. This isn't true, of course, since no heat is used. All we can say of the emitter (in the case of Fig. 3-17) is that it is electron-rich. The collector (a good name) is so called (in Fig. 3-17) because it collects electrons. In a tube, the equivalent would be the plate. We can consider the base in the manner of a control grid.

It would seem, in Fig. 3-18, that these names are now meaningless (except for the base) since the roles of collector and emitter are reversed. But this is only so because we have been considering current flow as electron flow. If, in Fig. 3-18, we think of the movement of holes, or positive charges, then the emitter is rich in positive charges and the collector is on the receiving end.

AMPLIFICATION

In the circuit of Fig. 3-17 we managed to transfer most of the current flowing in a low-resistance circuit (the emitter-base circuit) to a much higher one. Thinking of this from a resistance viewpoint, we transferred (changed) a low-resistance circuit to a high-resistance one. But from *trans*fer and res*istor* we manage to get transistor. Think of the transistor as being somewhat in the nature of a special type of stepup transformer. In this special transformer a variation of resistance in the primary will produce a variation of resistance in the secondary. It is this characteristic that enables us to use the transistor as an amplifier and an oscillator. Let's see how we can go about getting amplification. We can do this by examining the circuit in Fig. 3-19.

In Fig. 3-19 we have an npn transistor, a pair of bias batteries, a 300-ohm resistor in the emitter-base circuit (our input circuit) and a 5,100-ohm resistor in the collector-base circuit (our output circuit). Let us assume that we have a current of 3.33 mA beginning to flow. This current, in passing through the 300-ohm resistor, will produce a drop of almost 1 full volt across it. Most of the current (3 mA in this case) will pass through the base, over to the collector. A small amount of current (.33 mA) will go through the base to the input bias battery.

We now have 3 mA of collector current flowing through our

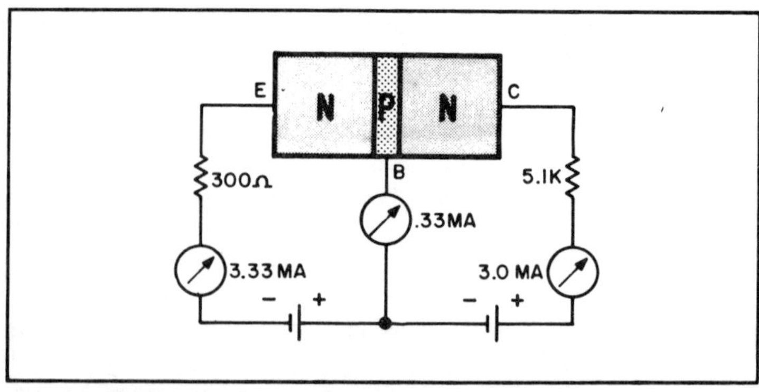

Fig. 3-19. Most of the current passes from the emitter to the collector. The remainder flows from the base to the forward-biasing battery.

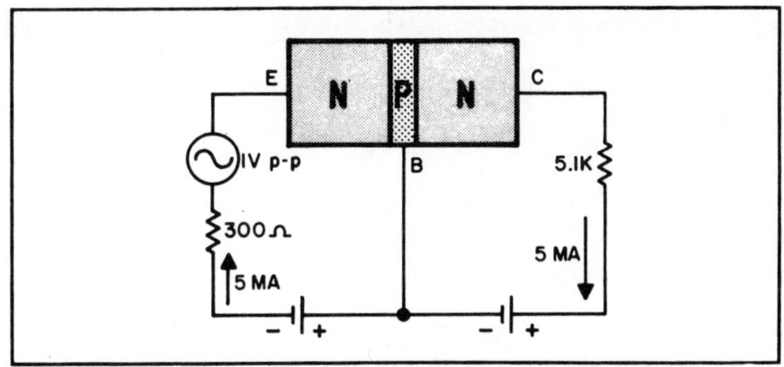

Fig. 3-20. A signal voltage (represented by the symbol for an ac generator) varies the amount of forward biasing, thus varying the current flowing to the collector.

5,100-ohm collector resistor. This will result in a drop of 15.3 volts across it.

So far we have nothing that could be called spectacular. The two resistors, plus the transistor, are acting as a dc voltage divider across the two series batteries. But suppose, as in Fig. 3-20, we put an ac signal source in series with the emitter-base input. This ac source is going to supply an emf that is 1 volt peak-to-peak. At its maximum negative peak point, the ac voltage will add one-half volt to the input battery voltage. At its maximum positive peak, it will subtract one-half volt. Consider what happens when one-half volt is added. Our forward-biased section becomes even more so. Since we have added a voltage in series aiding with our other batteries, more current will flow. In Fig. 3-20 the current will rise to 5 mA.

To make our arithmetic a bit easier, imagine that all of this current will go over to the collector. With 5 mA flowing through the 5,100-ohm resistor, the drop across it will now become 25.5 volts. But since we had 15.3 volts to begin with, we have an increase of 10.2 volts in our output circuit. What about the input? That went up by only one-half volt. Thus, we have a voltage gain in the ratio of 10.2 to 0.5, or 20.4 to 1.

When the signal voltage changes, it will be in opposition to the battery voltage. This will decrease the forward bias (or increase the input resistance), lowering the current. If we get a proportionate decrease in current, the ratio of output voltage to input voltage will still be 20.4 to 1.

We could have obtained exactly the same results by using a pnp transistor instead of the npn we used in Fig. 3-19. The batteries would be transposed. And while the voltage drops across the input and output resistors would still show us a ratio of 20.4 to 1, the

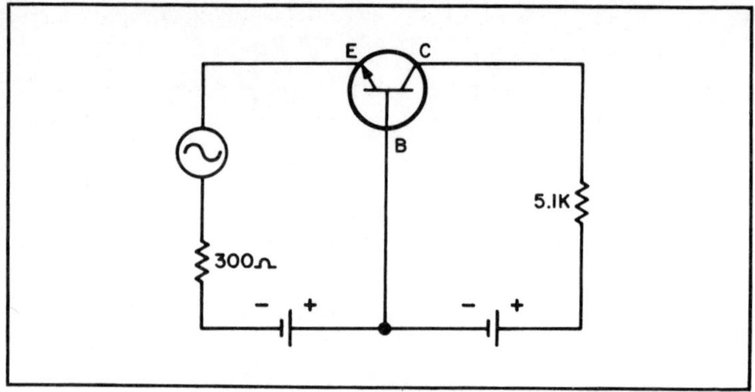

Fig. 3-21. Elementary transistor amplifier circuit.

polarities of these voltage drops would be exactly opposite to those of the npn circuit.

We know you are going to agree that the npn circuit of Fig. 3-20 can immediately be classified as a rather simple type. It certainly has few enough components! However, Fig. 3-20 is a sort of hybrid diagram, since we are using circuit elements for the components and a block-diagram style for the transistor. All we need to do now is take just one small step forward, use the symbol we learned earlier for the transistor, and end up with a complete schematic as shown in Fig. 3-21. This diagram, of course, represents the same action we got in Fig. 3-20.

CURRENT AMPLIFICATION—ALPHA

If, for some reason, we did not have the services of the forward-biased portion of the transistor (it could be shorted), collector current would be a function of the emitter-collector bias battery. Because of the reverse-biasing action, collector current would be very low indeed. However, if we allow our forward-biasing section to work properly, collector current rises considerably. In this sense, then, the transistor is a current-operated device.

Not all the electrons that enter the emitter make the full transit through the transistor. Percentage-wise, somewhere between 95 and 99 percent do manage to make the complete trip, but a few percent do get sidetracked through the base, back to the forward-biasing battery. If we manage to increase the emitter current (as we did before) the collector current will also increase (but not to the same extent). This ratio between a change in collector current and a change in emitter current is known as the current amplification factor. The Greek letter alpha (α) is used to indicate this characteristic of the transistor.

It may seem very strange that we talk about current amplification, especially since collector current is invariably lower than emitter current, but remember how this extraordinary situation came about. If it were not for the forward-biased emitter-base circuit, the current in the collector-base circuit would be a very small fraction of the current. Since alpha is less than unity, we generally discuss it in terms of percentage. Typical values of alpha are 95 to 99 percent.

CURRENT AMPLIFICATION—BETA

When we change the amount of emitter current, not only does the collector current change, but the base current as well. The ratio of the change in collector current to the change in base current can also be used as a measure of the amplification of the transistor. It is represented by the Greek letter beta (β). Since collector current is so much larger than base current, beta is in terms of whole numbers. Thus the beta of a transistor might be 20, 30 or 40, etc. A transistor with a beta of 40 means that 40 times more current flows in the collector than in the base circuit.

Quite naturally, the more current we have in our collector circuit the more current will flow through the resistor connected to the collector and the greater will be the voltage drop across it. Since our output is taken from across this resistor, we naturally want as much current through it as we can get. This means, in terms of alpha, that we would like to have alpha as high a percentage as possible—that is, as close to 100 percent as we can get. In terms of beta, we would want the number to be as large as possible. A transistor with a beta of 100 has more amplification than a transistor with a beta of only 20.

ALPHA AND BETA

Obviously, we cannot change alpha without affecting beta. The relationship between the two can be put into a very simple formula so that we can calculate the value of alpha (if we know beta) or the value of beta (if we know alpha). This is the formula:

$$\text{beta} = \text{alpha}/(1-\text{alpha})$$

We can make our work much easier if we use letters:

$$\beta = \alpha/(1-\alpha)$$

Example 1: A transistor has an alpha of 95 percent. What is the value of beta?

$$\beta = \alpha/(1-\alpha)$$

since 95 percent is the same as .95

$$\beta = .95/(1-.95)$$
$$\beta = .95/.05 = 19$$

Example 2: A transistor has a beta of 45. What is the value of alpha?

$$\beta = \alpha/(1-\alpha)$$
$$45 = \alpha/(1-\alpha)$$

If we multiply both sides of this equation by $1-\alpha$ we will get:

$$45(1-\alpha) = \alpha$$
$$45-45\alpha = \alpha$$

We can now transpose the -45α to the right-hand side:

$$45 = 45\alpha + \alpha \text{ or } 45 = 46\alpha$$

Dividing both sides of the equation by 46, we get:

$$45/46 = \alpha \text{ or } \alpha = .978 \text{ or } 97.8\%$$

VOLTAGE AMPLIFICATION

Ultimately, in dealing with an amplifier such as the transistor, we must come back to the idea of voltage amplification. We start with a signal voltage and this signal voltage changes the current flowing through the transistor. But this flow of current must be translated into a form we can use. In passing through the output resistor, the resistor connected in series between the collector and its bias battery, we get a voltage. To the extent that this varying voltage is greater than the signal voltage, we get amplification. The waveform of the voltage across the output resistor should be the same as that across the input resistor since the change in current is being produced by the input voltage. The technique is just the same as that used by a vacuum-tube amplifier where the signal voltage, inserted between grid and cathode, modifies the tube's bias. It might seem strange to talk about voltage amplification since the collector current is less than the emitter current, even though by just a small amount. It is the great difference between the output or load resistance and the input resistance that helps produce voltage gain. The difference between emitter and collector current might be small, but there is a deliberately large difference between input and output resistances.

THE TRANSISTOR AS AN OSCILLATOR

Basically, an oscillator is a dc to ac converter. We take the voltage and current supplied by the battery (or batteries) and change this dc voltage and current (with the help of the oscillator) into ac form. The most commonly used technique is to take some of the energy of the output circuit and feed it back, in phase, to the input. But before you decide that what we have there is a perpetual motion machine, remember that the energy is supplied by the battery, and that the ac power we get out of the oscillator is never quite as much as the dc power we put in. In oscillators, as in so many other things, you get nothing for nothing.

Figure 3-22 shows a simple oscillator arrangement. Oscillator circuits will be covered in greater detail in Chapter 6. What we have in Fig. 3-22 is a stripped-down version for purposes of explanation.

The only change of interest is the addition of the transformer which couples the output (collector) circuit back to the input (emitter). If you will trace the path of current flow, you will see that each transformer winding is in series with the transistor and the bias batteries.

The feedback winding, F, has very few turns. The resistance of the winding is very low and so its ohmic resistance has very little effect on the current. The winding soldered to the collector lead has many more turns. And while the current flowing through that wind-

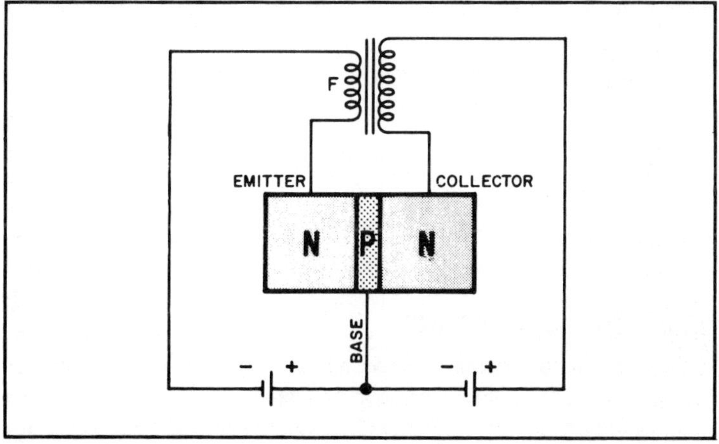

Fig. 3-22. Elementary transistor oscillator circuit.

ing is dc, yet the moment the circuit is closed, there is a rapidly growing magnetic field around the winding. A voltage is induced across winding F. But this voltage is in series with the emitter-base biasing voltage. If the induced voltage is phased so that it adds to the forward bias, the flow of current to the emitter will be increased. This increased current (most of it) will also flow through the collector and the collector winding. The growing magnetic field around the collector winding induces more voltage across winding F, which, in turn causes more current to flow.

But all good things must come to an end. The induced voltage across F depends on the rate of change of the magnetic flux. The combination of induced voltage plus battery bias is no guarantee that the current will continue to increase indefinitely. The current gradually slopes off and approaches a steady state value. All this means is that the current keeps increasing, but more and more slowly.

However, the change in the magnetic field around the collector winding depends on how rapidly the current through it changes. If the rate of current increase diminishes, so will the variation of the magnetic field around the collector winding. But this means less voltage will be induced across winding F. Less voltage, across F, though, means a smaller amount of current going to the emitter.

We now have a condition of a decreasing current through the collector winding. As a consequence, the polarity of the induced voltage changes, and now, the induced voltage, instead of aiding the forward bias battery, opposes it. With this sort of situation, the current rapidly drops to almost zero. This means that the magnetic field around the transformer is also so small as to be ineffectual. The forward bias battery now takes over again, because the opposition voltage across winding F has disappeared. This is where we came in, since the entire cycle of events repeats itself.

We have taken some liberties with this description of oscillator action since certain important components were omitted and since we have had no discussion as yet of transistor current curves. The description is of use since the behavior of a transistor as an oscillator is similar to that of its vacuum-tube counterpart.

TRANSISTOR MANUFACTURE

You will note that at no point in these first three chapters have we supplied any information on how transistors are manufactured. This is a highly specialized technique and there would be no point to our engaging in such discussion. However, there are numerous transistor types—so many that they deserve a chapter all to themselves. But even in the chapter on transistors (other than the triode) it would be impossible to describe all the new ones. Not only is there a large and growing variety, but developments now include semiconductors that behave somewhat like transistors, but which cannot be named as such. Fortunately, the basic theory presented earlier will help you reach a more ready understanding of the way in which they work.

So we come to the end of another chapter. We know you may be impatient to come to grips with practical circuits, but our introduction to the transistor has barely been completed. Before we concern ourselves with circuitry, we need to know more about the transistor, its special characteristics, its likes and dislikes, and its needs and its efficiency.

QUESTIONS

1. What is beta?

2. What is alpha?

3. A transistor has a beta of 65. What is the value of alpha?

4. A transistor has an alpha of 98 percent. What is the value of beta?

5. How do the values of emitter, base, and collector currents compare?

6. How is the emitter-base circuit biased? The collector-base circuit?

7. How does the volume of the base compare with that of the emitter or collector?

8. What is the technique used for varying the forward bias of a transistor?

9. How was the word transistor obtained?

10. What is the difference between high impedance and low impedance?

11. What is the difference between impedance and resistance?

12. What is reactance? What is the difference between reactance and resistance?

13. In a pnp transistor, which element is represented by the letter p? Which by the letter n? And which by the final letter p?

14. What is the difference in electron current flow in a pnp transistor compared to an npn type?

15. How does the input impedance of a transistor compare to the output impedance?

Chapter 4

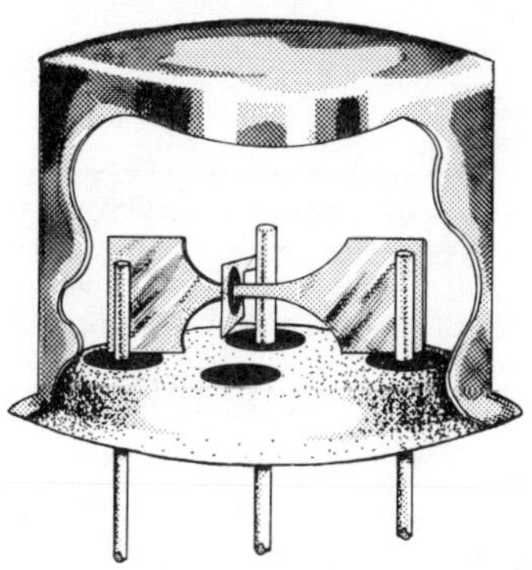

Transistor Characteristics

OUR INTRODUCTION TO THE TRANSISTOR IN THE LAST CHAPter was a bit formal, but now that the strangeness has worn off, we see that the transistor is just the application of some well-tested ideas but using a new technique. We have input and output circuits, just as with tubes. Our input is the base-emitter circuit, with the emitter (for the npn transistor) analogous to a cathode, and the base to a control grid. And, just as in the case of a tube, the input signal varies the bias. The output is the collector-emitter circuit, with the collector (again, for the npn type) reminiscent of the plate in a tube. (See Fig. 4-1).

Because you may be much more familiar with tubes than you are with transistors, the temptation to make comparisons between them will always be with you. In a way this will be unavoidable since many of our old, familiar tube circuits have been picked up, lock, stock, and resistor, and simply put to use with transistors . . . with some component changes. However, we just cannot pull a tube out of a circuit with the idea of substituting a transistor—and then sit back and hope for the best. If we do, the best will never arrive. A transistor has characteristics of its own.

BIASING METHODS

Transistors are much simpler than tubes in their voltage requirements. All that the triode transistor needs is a forward bias for

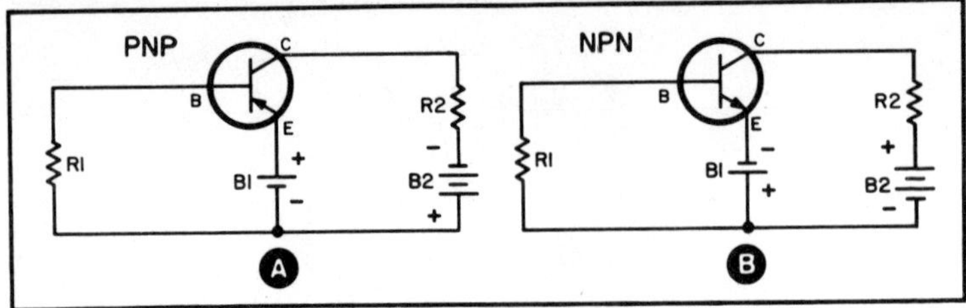

Fig. 4-1. The voltage between emitter and collector is the sum of the battery voltages of both pnp and npn circuits shown here. Although the circuits have been drawn somewhat differently than those in Chapter 3, if you will examine the circuits you will see that all biasing requirements have been met.

its base-emitter circuit and reverse bias for its collector-emitter circuit.

Figure 4-1 shows one possible arrangement for bias for a pnp and an npn transistor. The voltage between collector and emitter is the sum of the voltages of B1 and B2 just as it was in the circuits you saw in the last chapter. In the new arrangement of Fig. 4-1, though, both the input and output circuits share the voltage of B1. We can re-arrange the batteries as shown in Fig. 4-2 so that each circuit—input and output—has its own bias supply. However, in Fig. 4-2, the voltage between collector and emitter is now reduced by the amount of B1. We can overcome this by increasing the value of B2. The disadvantage of the circuit of Fig. 4-1 is that we cannot change the forward bias without, at the same time, changing the reverse bias. The arrangement in Fig. 4-2 eliminates this difficulty, but we do lose the advantage of the added voltage supplied by the series connection of Fig. 4-1.

SINGLE BATTERY OPERATION

We can eliminate one of the batteries altogether, as shown in Fig. 4-3. Note that in this npn circuit, the collector is still positive

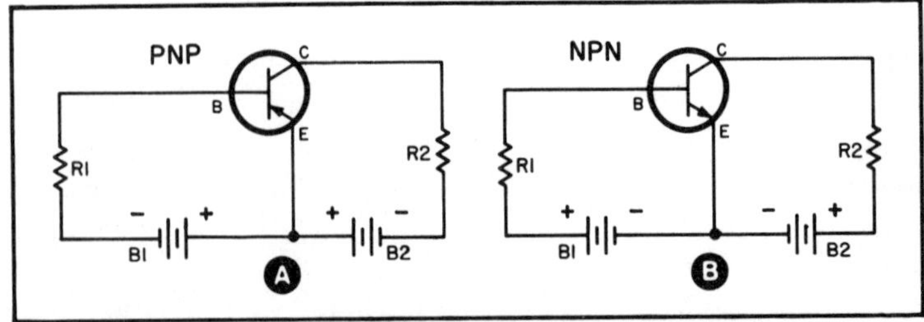

Fig. 4-2. In these circuit arrangements, the bias voltages are independent of each other, but now battery B1 does not add to the voltage between emitter and collector.

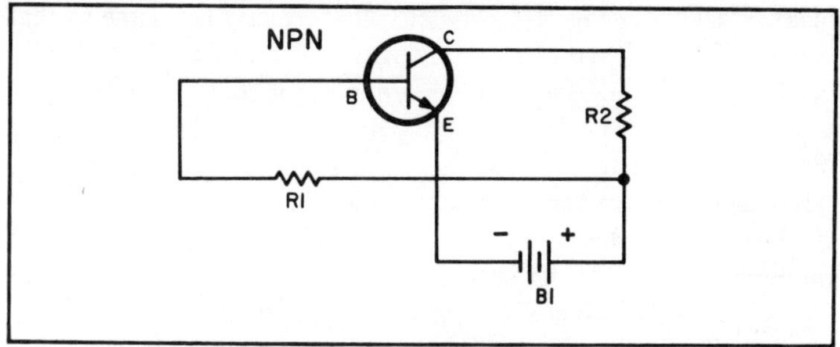

Fig. 4-3. A single battery, B1, supplies both forward and reverse bias. The correct amounts of bias can be obtained by selecting the proper values for R1 and R2.

with respect to the emitter, while the base is also positive. Thus, we meet the conditions for biasing this transistor. Now it might seem odd that we can do this, since the input and output circuits require different amounts of biasing voltage. But you will note that R1 is in series with the battery in the base-emitter circuit while R2 is in series with the same battery in the collector-emitter circuit. By selecting the right values for R1 and R2, we can still have the voltages we want. This might seem to give us the flexibility of the circuit of Fig. 4-2—and so it does—but remember that we pay for it by having voltage drops across R1 and R2.

FIXED BIAS

Although the circuit of Fig. 4-1, 4-2, and 4-3 are different, they do have one item in common. The *forward bias* has a *fixed* value, the amount of bias being determined by the voltage of the forward biasing battery (B1) and the value of R1. In these circuits, to change the forward bias we would generally use a different battery. For a particular value of base bias, the bias current remains constant. Fixed bias has the serious disadvantage that it is best for one particular operating point or set of conditions only. There are two factors that make life difficult for fixed bias. Transistors are temperature-sensitive. This means that the operating characteristics of a transistor will change with temperature. Also, there is quite a bit of variation, even between transistors of the same type number, so that if you replace a transistor (in a circuit having fixed bias) you cannot be sure that the bias is correct. To make it correct, you would need to experiment with B1, or R1, or both.

SELF BIAS

We can make the bias self-adjusting (which is why we call it self bias) as shown in Fig. 4-4. The circuit is exactly the same as that shown in Fig. 4-3, with but one change. In both cases, one end of R1 is connected to the base. The other end of R1, though, (in Fig. 4-4) is wired to the top end of R2. Now let's see how this changes the circuit from fixed to self bias. First, using Fig. 4-3 as our guide, let's trace the path of the base-emitter circuit. Starting at the emitter, we go through the transistor to the base, from there through R1, then through battery B1, and so back to the emitter. But if we try to follow the same path in Fig. 4-4, we now find that we must include the collector load resistor R2, in our path. What difference can this possibly make?

In Fig. 4-4 we have an arrow immediately next to R2. This arrow shows the direction of current flow and the polarity of the voltage drop, across R2, due to collector current. This voltage is in opposition to the battery voltage. But the voltage between the base and the emitter (our forward bias) depends almost entirely on the arithmetic sum of these two voltages—the battery voltage and the drop across R2. If collector current is large, so is the voltage across R2. A large voltage across R2, though, means less forward bias. If forward bias decreases, less current passes through the transistor and so collector current decreases. This reduces the drop across R2. Less voltage across R2 means more forward bias.

Sometimes the act of drawing a circuit hides the information it is supposed to convey. Generally, most circuits are prepared so that there is an orderly progression of signal, from left to right. This means that we start with a weak signal at the input, moving through circuit after circuit, toward a strong signal at the output. In a way this makes sense since we then try to "read" the circuit just as though we were trying to read a line. Its disadvantage is that the circuit becomes less clear electronically.

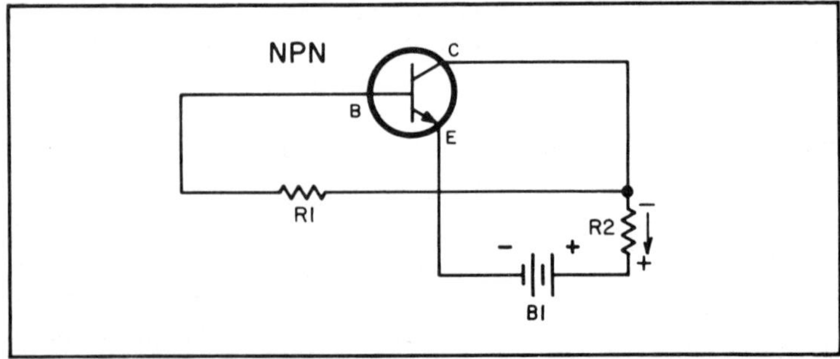

Fig. 4-4. In this circuit, forward biasing depends on the amount of collector current flowing through R2.

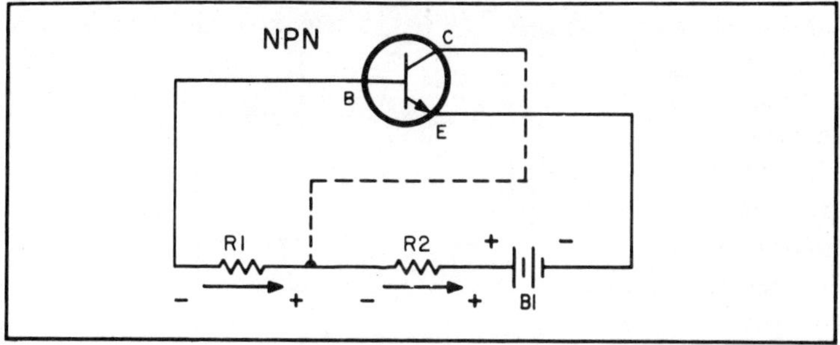

Fig. 4-5. This is not the way in which we usually draw a transistor circuit, but it does show forward biasing very clearly.

To understand what this is all about, take another look at Fig. 4-4. To understand what is happening in the way of forward biasing, you would need to trace the entire circuit between the input elements—the base and the emitter in this case. While this might be fairly easy to do in the case of Fig. 4-4, remember that what we are dealing with here is a very simple circuit. It might not be so simple when associated with a half-dozen or more transistors and assorted parts.

To see just what is happening in the way of bias, examine Fig. 4-5. Here we have redrawn Fig. 4-4 in a somewhat unorthodox manner so that you can see what is happening quite clearly. Let's trace the paths of current flow, beginning at the negative terminal of B1. From here we move into the transistor, through the emitter and base, through R1 and R2, and back to the battery once again.

The arrows across R1 and R2 show the direction of electron flow and because we know this direction, we know the polarity of the voltage drops across R1 and R2. These voltage drops are in opposition to the battery voltage. And what is the voltage between emitter and base—in other words, what is the amount of forward bias? It is equal to the voltage supplied by battery B1 *minus* the voltage drops across the two resistors—R1 and R2.

Now let's consider R1 and R2. Because we have drawn them with equal physical dimensions, we might succumb to the temptation to regard them as having equal voltage drops. There are several reasons why this is most definitely not so. First, consider the relative resistances of R1 and R2. R1 has a very small value, much smaller than that of R2. But there is another important consideration. Examine the dashed line we have drawn from the collector to the junction of the two resistors. This represents another path of current flow—a path that does *not* include R1. Furthermore, just to

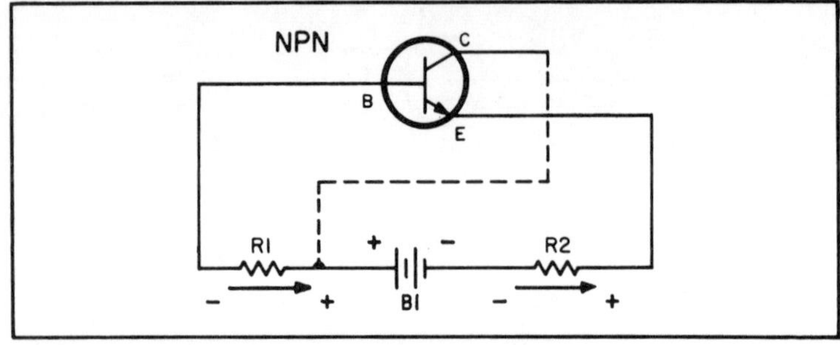

Fig. 4-6. Although B1 and R2 have been transposed (see Fig. 4-5), circuit operation remains exactly the same.

make the evidence more compelling, this current is many times greater than the current flowing through R1. All in all, the voltage drop across R2 is much greater than that across R1 and for this reason is the determining factor in the forward biasing of the transistor.

One other fact emerges from a study of Fig. 4-5. We can transpose R2 and B1 without affecting the circuit. When we do this the amount of current flow through R1 and R2 is not changed. If you will examine Fig. 4-6 you will see that the status of the components and their behavior has not been altered. But if you will now move along to Fig. 4-7, you might very well imagine that we have a completely new circuit. Actually, it is not new in the slightest. It is identical with that of Fig. 4-4. The difference (as in the case of Figs. 4-5 and 4-6) is in the transposition of B1 and R2, a change that does not affect operation of the circuit.

We've had a rather steady diet of resistors in the output, so

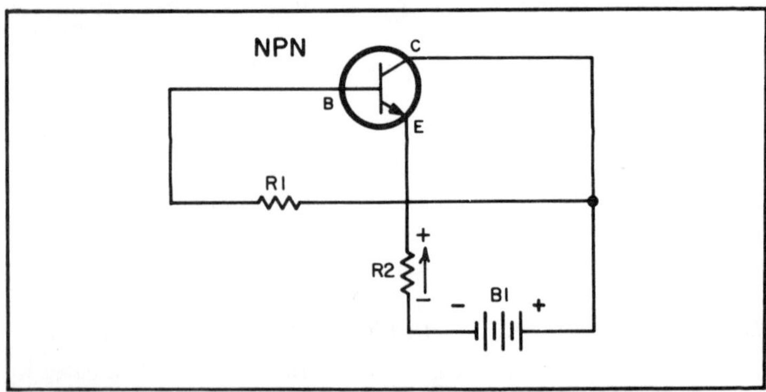

Fig. 4-7. This circuit is identical with that of Fig. 4-4. Transposition of B1 and R2 has not changed circuit operation.

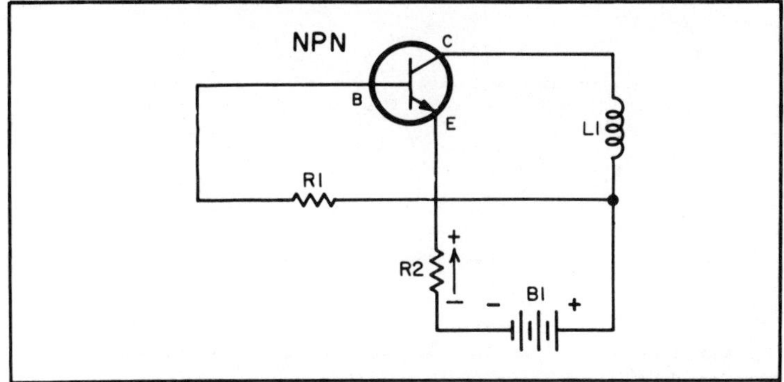

Fig. 4-8. In this circuit, coil L1 is used instead of a load resistor. This coil could be the primary of a transformer.

suppose we insert a coil, L1, as shown in Fig. 4-8. But because we want variety, we have given ourselves a problem. We had depended on the resistance of the load to give us a voltage drop. Our coil, though, isn't so obliging. We can solve this difficulty quite easily by inserting a resistor, R2, in the emitter circuit. This resistor performs exactly in the same way (as far as forward bias is concerned) as load resistor R2 in Fig. 4-4. The arrow alongside R2 gives us the direction of electron current flow and also the polarity of the voltage drop across R2. Again, this voltage is in opposition to the battery voltage. If you will take a walk, starting at the emitter, through the transistor to the base, you will see that we can continue through R1, the battery, and then R2. The position of the resistor hasn't changed the fact that we still have self bias.

COLLECTOR VOLTAGE VERSUS COLLECTOR CURRENT

To learn the condition of a patient, a doctor pokes, probes, listens with his stethoscope, and finally gets a picture of the patient's general state of health. We follow a similar routine with transistors, except that our equipment consists of a microammeter (to measure base current) and a milliammeter (to measure collector current) as shown in Fig. 4-9. While this setup is for an npn transistor, the same arrangement (with transposed batteries) would do for a pnp.

COLLECTOR CHARACTERISTIC CURVE

With the test circuit of Fig. 4-9 we can obtain a relationship between collector voltage and collector current. This is a static test. All that this means is that we are going to apply dc voltages

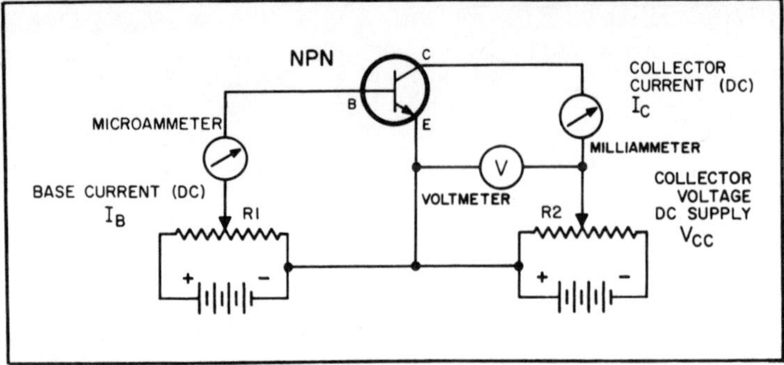

Fig. 4-9. Setup for determining the relationship between collector voltage and collector current.

only. You might think that this isn't quite fair, since our transistor is ultimately going to work with a signal input, yet the collector characteristic is important. We've been talking about putting bias voltages on our transistor, and have examined several circuits. The collector characteristic is going to help us come to a decision on just what values of bias we should use and what sort of a resistor we should have as our collector load.

Figure 4-10 shows the results of one test using the circuit of Fig. 4-9. R2 was set for different values of collector voltage. R1 was adjusted to keep the base current at a constant 40 microamperes.

Now examine the curve between the points at which we have 2 volts on the collector, and 6 volts. At 2 volts, the collector current is

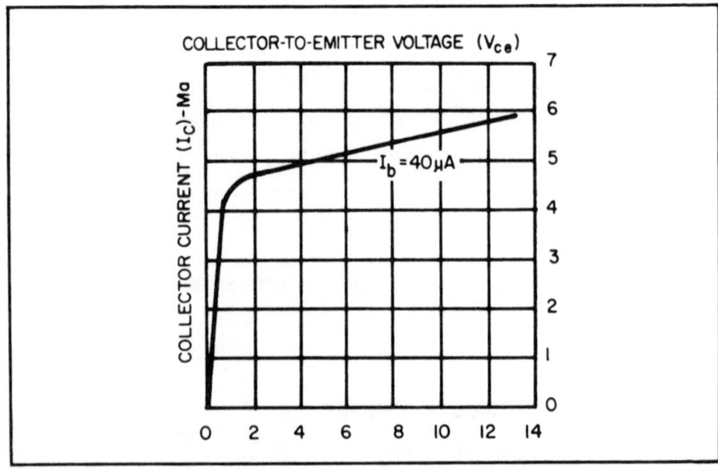

Fig. 4-10. Collector characteristic curve. Note how little the collector current changes with collector voltage for a given value of base current.

slightly less than 5 mA. At 6 volts it is slightly above 5 mA. This is quite a range of collector voltage change, and yet the collector current has remained substantially constant, with a constant amount of base current.

If we were testing a tube, we would be measuring plate voltage versus plate current, for a particular value of grid bias and in this way we would obtain a plate characteristic curve. In Fig. 4-9 we measured collector voltage versus collector current for a fixed amount of bias current, and obtained a collector characteristic curve. Thus, base current in the transistor performs the same function as grid bias voltage in a tube.

TUBE VERSUS TRANSISTOR BIASING

There is, of course, a mental hazard in equating vacuum-tube grid bias and the technique we use for forward-biasing a transistor. In the vacuum tube, an increase in bias means that we are making the control grid more negative with respect to the cathode. As far as bias is concerned, we are moving in a negative direction and the result is a decrease in plate current. In the case of the transistor, the base is biased either negative or positive with respect to the emitter, depending on whether we are using a pnp or npn type. Furthermore, an increase in bias, regardless of the transistor type, means an increase in collector current, an action quite opposite to the plate current decrease resulting from raising the bias in a vacuum tube.

COLLECTOR CHARACTERISTIC FAMILY

A single collector characteristic, such as that shown in Fig. 4-10, supplies limited information. To learn more about the transistor, we can have a whole family of collector characteristics, as in Fig. 4-11. The procedure for obtaining these curves is exactly the same as that outlined earlier. In this we get a whole family of curves, all the way from zero base current, in steps of 10 microamperes, up to 60 microamperes. Note how similar the collector characteristic curves are to the plate characteristics for a 6AK6 pentode, shown in Fig. 4-12. The curves are similar, yet the transistor we have been testing is a triode. The straight line portion of the curve, (whether tube or transistor) is the useful part. In the transistor, we get to the straight line portion quite rapidly. In Fig. 4-11, all we need is about one quarter to one half volt. For the tube, we need about 75 volts on the plate before we get the same results.

This gives us another clue to transistor versus tube behavior.

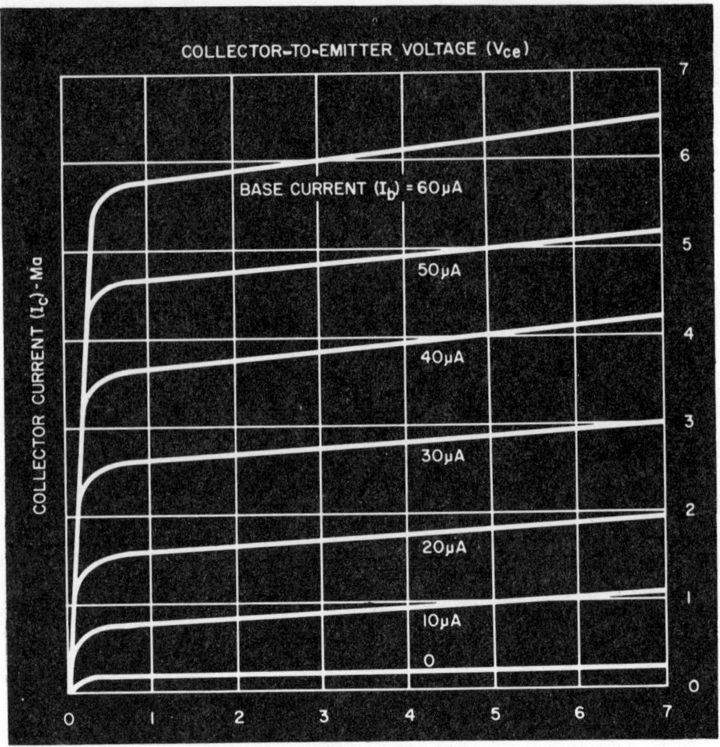

Fig. 4-11. Family of characteristic curves for a transistor.

Transistors not only work with smaller voltages than tubes, but respond to much smaller *variations* in voltage. If you're accustomed to tube potentials, revise your thinking downward for transistors.

THE COLLECTOR LOAD

In a transistor circuit (as in a tube circuit) the components must work together as a team. It takes just one wrong part or value to make the circuit act sluggishly, or not at all. This doesn't mean that there is only one right combination of parts and values —there are many. The whole idea is to select the one right group out of the many available.

At first thought this might seem like an impossible suggestion, since the number of variables involved would give us a very large number of component combinations. That is why characteristic curves of transistors—and of tubes—are so important. They give us some idea about the behavior of these components and enable us to

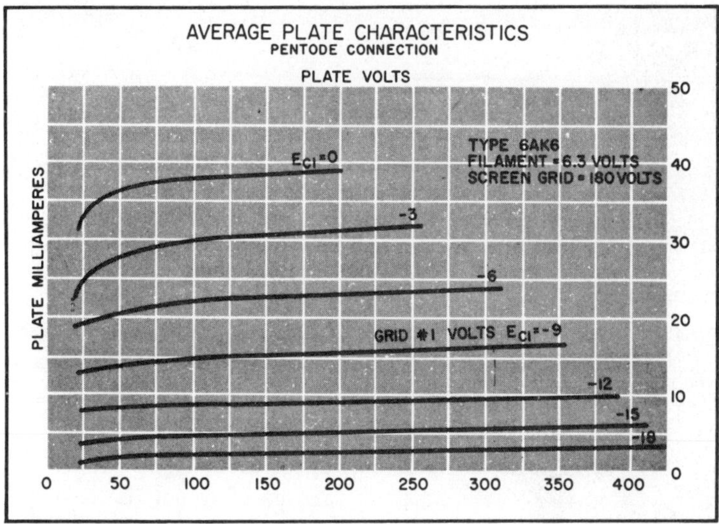

Fig. 4-12. Pentode characteristic curves. These are similar to characteristic curves for a transistor. The transistor, though, is a triode while the curves shown here are for a pentode tube.

make a selection of associated parts in a more intelligent manner.

To see how this concept works, start with Fig. 4-13 which gives the collector characteristics for a transistor. Since the collector-to-emitter voltage ranges from zero to 20, let's start with

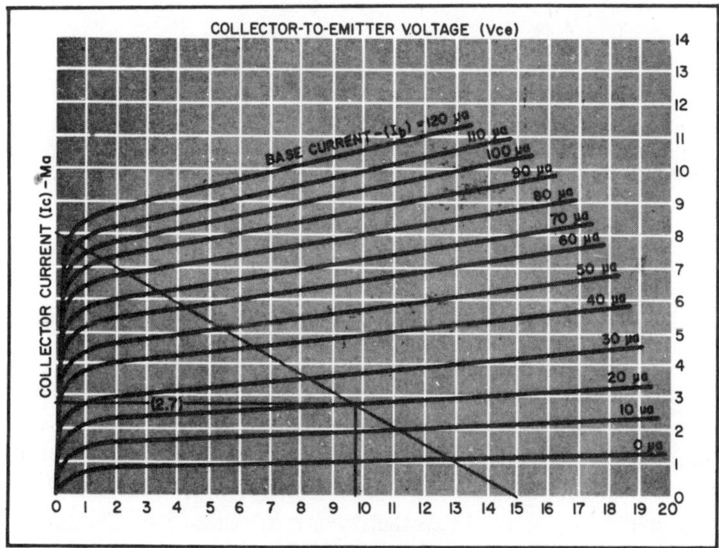

Fig. 4-13. Family of collector characteristics.

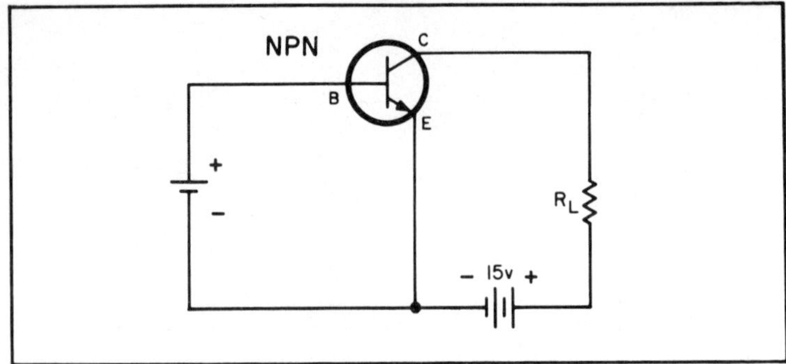

Fig. 4-14. Simple circuit used for measuring collector characteristics. The value of forward bias can be changed to obtain the base current values shown in Fig. 4-13.

a battery of 15 volts, using the circuit shown in Fig. 4-14. Why 15 volts? It is just a starting point, and possibly not the best one either.

Using the circuit of Fig. 4-14, suppose we connect a wire between the collector aand emitter. Since this is a short across these two elements of the transistor, the collector to emitter voltage becomes zero. The full voltage of the battery is now across the load resistor, R_L. The amount of current will now depend only on the battery voltage and the value of R_L. Suppose this current is 8 milliamperes. We now have a condition of:

collector-to-emitter volts = 0
current through R_L = 8 mA

At the left side of the graph (Fig. 4-13) find the zero point. Now move up vertically until you reach the 8 mA point and put a dot right there. Now draw a straight line between this point and the 15 volt point on the base line, just as shown in Fig. 4-13. This is our load line. We can find its value by using Ohm's law:

$$R = E/I = 15/.008 = 1{,}875 \text{ ohms}$$

We cannot dismiss the graph in Fig. 4-13 now because it contains information we need. Let's move along, first, to the 15-volt point on the base line. What is this 15 volts? We know it is our full battery voltage, but there is more to it than just that. The description on our base line tells us that we are working with collector-to-emitter voltage—that is, the voltage existing between the collector and the emitter. But for a full 15 volts to exist between collector and emitter, we cannot afford any voltage drop across R_L. We can eliminate the drop across R_L in two ways. We can assume the transistor is at cutoff—that is, no collector current flowing—or else we can put a shorting wire right across R_L. Whichever method we use, the voltage between collector and emitter would be the full battery voltage.

However, we did put the load resistor in the circuit for a purpose—to develop an output signal across it, so now let us remove the short from across R_L or else take the transistor out of cutoff. If collector current flows at this time, it will also pass through R_L, producing a voltage drop across it. This means we will now have less voltage between collector and emitter. Thus, if we have a 5-volt drop across R_L we will have:

$$15 - 5 = 10 \text{ volts between collector and emitter}$$

At this time, how much current is flowing in the output circuit? Move your finger along the base line until you reach the 10-volt point. Now move straight up until you reach the load line. Then move across horizontally, toward the left, until you reach the vertical current line. The intercept is at 2.7 milliamperes.

What do we learn from this? As the voltage between collector and emitter becomes smaller, current in the output circuit gets larger. Carrying this to its logical conclusion, our graph shows us that when the voltage between collector and emitter becomes zero, we have a maximum current of 8 mA flowing through the load.

We need only two points to determine the load line. We used two extreme conditions:—1) a condition of maximum current (8 mA) and zero collector-to-emitter voltage and 2) a condition of maximum voltage (15 volts) and zero current in the output circuit.

A little earlier we found the value of the load line by using maximum values, but this value is confirmed even when current is flowing through both transistor and its load. In Fig. 4-13 we see that for a 5-volt drop across the load, we have 10 volts remaining between collector and emitter. For this value of voltage, we get a current of 2.7 mA. The load value then is:

$$R_L = \frac{5 \text{ volts}}{.0027 \text{ ampere}} = 1{,}852 \text{ ohms}$$

Obviously, the value of the load is equal to the voltage across it divided by the current through it. The graph in Fig. 4-13 gives us this information. The voltage across the load is equal to the battery voltage minus the drop between collector and emitter.

OTHER VALUES OF LOAD

The slope of our load line depends on the battery voltage and the current that flows in the circuit with the collector-emitter shorted. The slope becomes steeper for different values of load as shown in Fig. 4-15.

We can use load lines to see which one will give us the least amount of distortion. Suppose we assume that we are going to start with a base current of 45 microamperes. We will then decrease our forward bias, dropping the base current to 30 microamperes, and then increase it to 60 microamperes; that is, we will decrease and increase our base current by 15 microamperes. When we do, and tabulate our results, this is what we will get:

When R_L = 3,333 ohms

base current	collector current	difference
30 μA	3.6 mA	
		1.2 mA
45 μA	4.8 mA	
		1.0 mA
60 μA	5.8 mA	

When R_L = 2,500 ohms

base current	collector current	difference
30 μA	3.9 mA	
		1.1 mA
45 μA	5.0 mA	
		1.2 mA
60 μA	6.2 mA	

When R_L = 2,000 ohms

base current	collector current	difference
30 μA	3.9 mA	
		1.2 mA
45 μA	5.1 mA	
		1.3 mA
60 μA	6.4 mA	

Since we get almost equal swings of collector current for equal changes of base current, distortion in the output will be small. Note that our operating point (sometimes called the quiescent or Q point) is at the intersection of the load line and the base current curve.

How do we know that we won't get distortion? Look at the base current curves. They are fairly equally spaced. This means that an increase or decrease of base current will produce fairly equal changes in collector current. If the base current lines were not equally spaced, we would not get equal changes in collector current for equal changes in base current. You can see an example of this sad

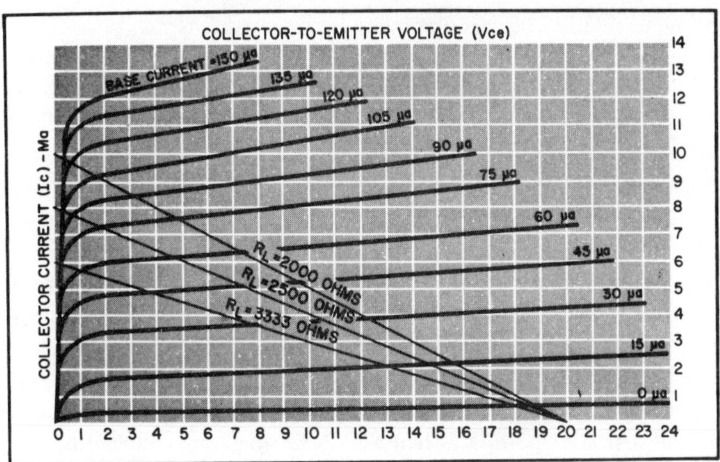

Fig. 4-15. When the load line intersects base-current curves that are equally spaced, distortion produced by the transistor will be at a minimum.

situation in Fig. 4-16. Our operating point is at the intersection of the load line with the curve for a base current of 200 microamperes. Our input signal has a peak-to-peak value of 200 microamperes—that is, it has a positive component of 100 microamperes and a negative component of the same amount. Our base current, then, is

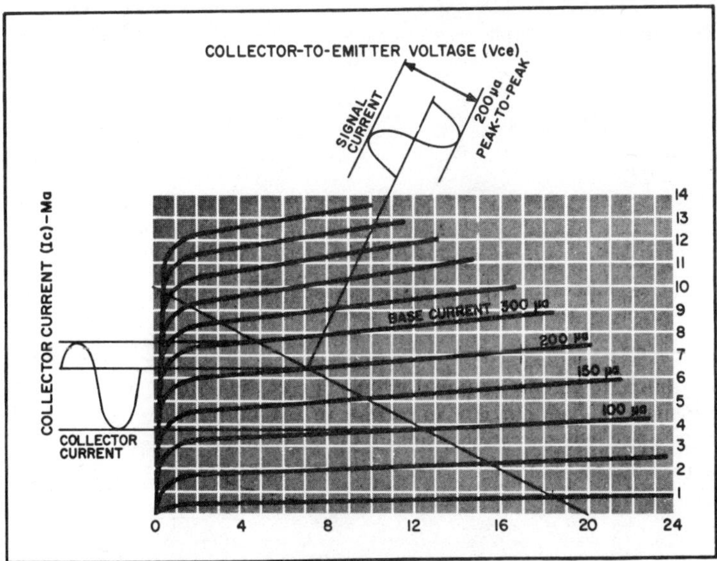

Fig. 4-16. Distortion will result because of the unequal spacing of the base-current curves above and below the quiescent point.

going to swing 100 microamperes either way. Note the uneven spacing of the base current curves. The distance from our starting point to the 300 microampere curve is much smaller than the distance from the starting point to the 100 microampere curve. This means that our collector current variation is not going to be symmetrical. Our input signal current, though, is symmetrical. What the graph is trying to tell us is that our transistor is going to distort the signal.

How serious is the situation in Fig. 4-16? The collector current waveform doesn't look too different than the signal current waveform. The answer here is that the amplitudes of the two waveforms mean little until we consider that the signal current is in microamperes and the output or collector current is in milliamperes. A milliampere is a thousand microamperes. Viewed in this light, if we consider the signal current as "same size," then the collector current waveform is a thousand times larger. Does this mean that the gain of our transistor is a thousand? Hardly! The variation in output voltage is still the product of the collector current multiplied by the value of the resistance through which it flows.

MUTUAL OR TRANSFER CHARACTERISTICS

The graph in Fig. 4-16 is a bit of scientific fortune-telling. With its help, we can begin to predict just how the transistor will behave and what we can expect from it. But you can't expect a transistor to reveal all in just one picture, so, just to be able to say that we are more than just nodding acquaintances, let's work with a new graph of transistor characteristics—the one shown in Fig. 4-17.

In Fig. 4-17 we have plotted (sounds sinister, but it isn't) base-to-emitter voltage versus collector current. But what is base-to-emitter voltage? Just our old friend, forward bias. And, as the curves show, as our forward bias goes up, so does our collector current.

There are a few exciting facts we can gather here. Take a look at the lowest curve—the one marked 70° C. When the forward bias is zero, so is the collector current. But as we increase the forward bias—not by volts, mind you—but by tiny fractions of a volt—whoosh, away goes the collector current.

There's still another item that's just as important. As the temperature changes, so does the collector current. Actually, we should have had just one curve. Having the temperature affect the collector current isn't at all desirable. In Fig. 4-17 we have only

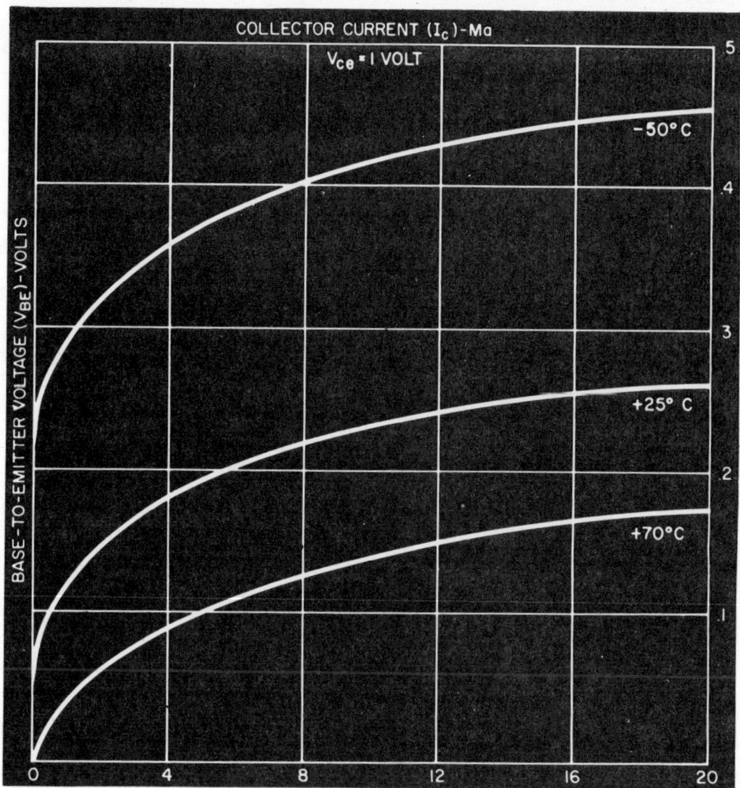

Fig. 4-17. Transfer characteristic curves. These graphs show the effect of temperature on collector current.

three curves because the temperature changed only three times. But what if the temperature kept going up and down. Which curve would be the right one? You can be sure we're going to do something about that, in just a few moments.

In the meantime, why should we consider a graph of the type in Fig. 4-17 so important? This graph gives us an insight into the amplifying properties of the transistor. We can see (we don't have to imagine) the effect of forward bias on the output or collector current. Curves of this sort are known as transfer or mutual characteristic curves.

STABILIZATION

Most tubes (there are some exceptions) operate at respectably high temperatures. Try removing a power tube (after it has been on for a few minutes) with your bare fingers, and you'll have a dramatic example of what we mean. Even low-voltage miniatures have a temperature much higher than that of the surrounding air. This means that tubes aren't affected by room temperature—they have their own, warm operating region.

But transistors don't have heated filaments or cathodes. They don't have plates that must dissipate heat. It doesn't take much of a rise in temperature to set the electrons in a bit of n-type germanium or silicon to thinking about traveling over to those very atttractive positive charges just on the other side of the potential barrier. But in a transistor (or in a tube) it is we who must control the current, and not temperature, or any other nonsense of that sort.

You might think that we're making a big fuss over a small matter, but in Fig. 4-17 we went to the trouble to draw graphs of transistor behavior for you. Temperature changes mean that these curves vary all over the place. And if these curves change then so will the current gain change, and naturally, right along with it, the output impedance will vary. This is definitely not a good situation.

We can "immunize" the transistor, protect it against the seductive effects of temperature by stabilizing it. One technique we can use, known as current feedback, is shown in Fig. 4-18. The circuit is similar to one we talked about earlier—Fig. 4-17. However, let's consider R2 from a new point of view.

R2 is in a rather unique position. Trace the output circuit and you will see that R2 is part of it. But you can do the same with the input circuit. R2, then, is common to both, and holding on to this tidbit of information in our hot little hands, we can learn of two ways in which R2 overcomes the effects of temperature.

Let us say that the temperature does change and in such a way that the collector current increases. What other change could have produced the same result? We could have increased the collector voltage or we could have increased the forward bias, or we

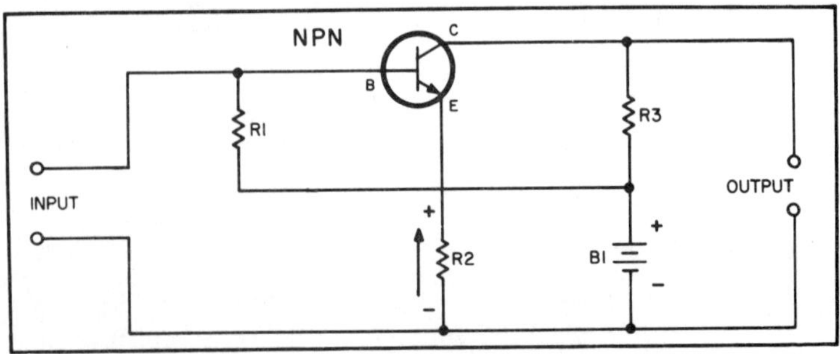

Fig. 4-18. R2, common to both input and output circuits, is a stabilizing resistor.

could have done a little bit of both. But, as far as we know in this case, none of these things happened.

With an increase in temperature, our collector current starts to rise. The collector current not only flows through R3, but through R2 as well. Note the direction of current flow through R2. This produces a voltage drop across R2 whose polarity is such that the top end of R2 (the end connected to the emitter) is plus and the bottom end (the end connected to the base through the signal input) is minus. Since our transistor is an npn type, our forward bias works by putting a negative voltage on the emitter and a positive voltage on the base. The voltage across R2, though, acts as a reverse bias for the base-emitter circuit. Or, we could say that the forward bias is being opposed by the voltage across R2. Since, in effect, the forward bias is being reduced, the collector current decreases.

This isn't the only action that takes place. Collector current depends on the voltage existing between collector and emitter. This voltage is supplied by the battery. The actual voltage between collector and emitter, though, isn't the full amount of battery voltage for we must subtract the amount of voltage dropped across load resistor R3. In addition, in the case of the circuit of Fig. 4-18, we must also subtract the voltage drop across R2.

Thus, R2 acts in two ways to counteract the increase in collector current caused by a rise in temperature. It reduces forward bias and it also reduces the collector-emitter voltage.

CURRENT FEEDBACK

Components in some circuits often perform a dual function, but the name given to the component is often a concession to the more important function. Thus, a coupling capacitor is so called because it couples one stage to the next, but, at the same time, it is a blocking capacitor, preventing the dc voltage of a previous stage from moving over into the following one.

In Fig. 4-18, R2 is a stabilizing resistor, but it does more than just that. Let us imagine that we now have an incoming signal. If the signal (at a particular moment) increases the forward bias, collector current will increase. But this increase in collector current will produce more voltage across R2, which, of course, has a certain amount of opposition to the increase in forward bias. Since the voltage across R2 will vary at the same rate as the input signal, but will oppose it, we can see that we have an out-of-phase condition, or a form of negative feedback. And, since negative feedback reduces the gain of the stage, that is exactly the sort of result we can expect from R2. This type of negative feedback is called current feedback.

Negative feedback is a sensible way in which to stabilize a transistor against increases in current due to temperature rises. It's just like having too much water pouring out of a faucet. We don't telephone back to the reservoir and tell them to cut down on the pressure. All we do is turn the valve a bit.

Since negative feedback is the solution to our temperature problem, we can use any form of out-of-phase feedback (negative feedback) we want. Figure 4-19 is an example of what we mean. Here we have an npn transistor. As we make the base more positive (a positive-going signal can do that for us), collector current increases. This produces a voltage drop across R3. The arrow shows us that the top end of R3 is negative-going (it is becoming more negative with respect to its other end). All we need do now, is connect the negative-going end, through R4, to our positive-going base. This is a form of negative feedback, since the voltage R4 steals from R3 is out of phase with the input signal. We call this voltage feedback.

If you've examined the circuit of Fig. 4-19 carefully, you've probably observed that we still have R2 in the circuit. Shunted across R2 is an electrolytic capacitor. We could remove this capacitor, and then we would have both voltage feedback (supplied by R4) and current feedback (supplied by R2). Feedback of this sort not only keeps temperature changes at arm's length, but also improves the performance of the circuit. We're not going to discuss it further at this point, since we would be encroaching on the material in Chapter 5.

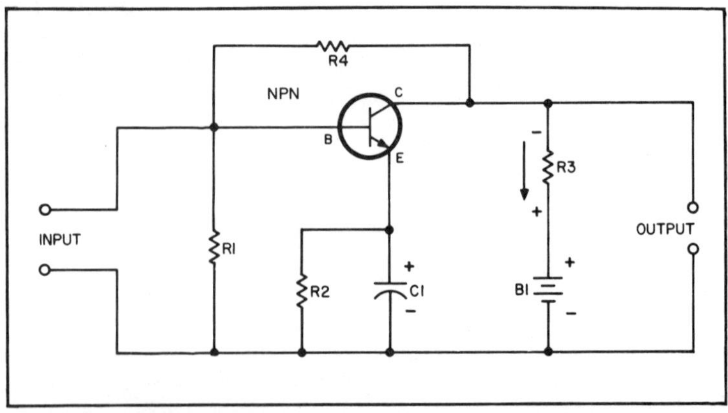

Fig. 4-19. We can immunize the transistor against the effects of temperature by using voltage and current feedback.

INPUT RESISTANCE

Whenever we have voltage *and* current, we always have resistance. We may not even have a physical resistor in the circuit, but the fact that we can divide the voltage across a pair of points by the current flowing between those points, means that we must have resistance. And the movement of current is one of those facts of life we must always associate with the transistor. Whether or not we have a signal, we always have current flowing in both input and output circuits.

Resistance can be measured directly or indirectly. We can put the leads of an ohmmeter right across a resistor, and there you are—the value of resistance can be read directly from a scale on the meter. But if we have current flowing (whether or not we have an actual resistor in the circuit) we must put our ohmmeter away and measure resistance indirectly.

Figure 4-20 shows how easy this is to do. Here we have a dc microammeter connected between base and emitter. This will measure our base current. And, shunted across the base and the emitter we have a high-impedance vacuum-tube voltmeter to measure our voltage. Divide the voltage reading by the current reading and we have the input resistance.

In Fig. 4-20 the collector floats—that is, we have no provision for reverse bias. The collector-emitter is open circuited. Remember—a transistor is different from a tube in that there is no isolation between the elements. The emitter, base, and collector are in close contact. Thus, in measuring the input resistance we want nothing in the collector circuit to affect our test.

What value of voltage shall we use in Fig. 4-20? We use the amount of forward bias voltage determined by the location of our

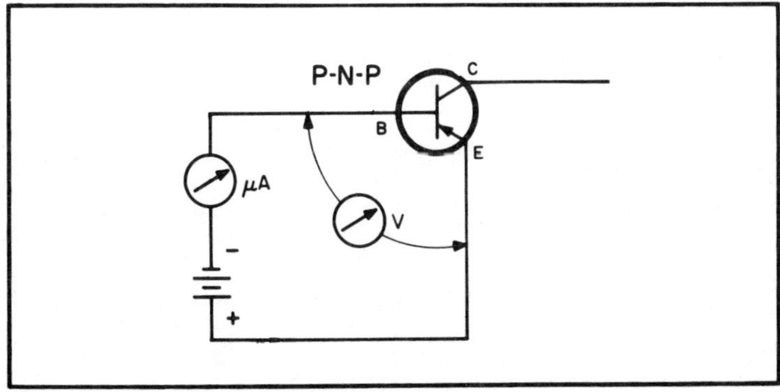

Fig. 4-20. Setup for measuring input resistance.

operating point on the load line. This value is used since this is the value at which the transistor will operate.

OUTPUT RESISTANCE

In measuring output resistance, our input will now be open circuited, but the transistor will have its operating reverse bias applied, as shown in Fig. 4-21. And, while we are on the subject of input and output resistance, keep in mind that we can run the same tests on npn units just as easily (Figs. 4-20 and 4-21 show pnps). All we do is transpose everything—the leads to our test meters and the leads to our battery.

SMALL AND LARGE POWER

In a vacuum-tube circuit, events move along very smoothly until we reach the speaker. Most speakers are current-operated devices. This means that the tube—or transistor—preceding the speaker must be a power type. All tubes preceding the output tube can be voltage amplifiers. Voltage amplification is their primary concern. But the output tube (because of the demands of the speaker) is interested in power—the product of volts and current.

Transistors are not excused from these requirements just because they are transistors. Unlike tubes, transistors are current operated devices—that is, current flows in both the input and output circuits of a transistor at all times. But how much current? For transistors preceding the output transistor, currents are measured in microamperes and milliamperes. The output transistor, though, like its tube counterpart, must be a power type. In its output circuit we will have current in the order of milliamperes or amperes.

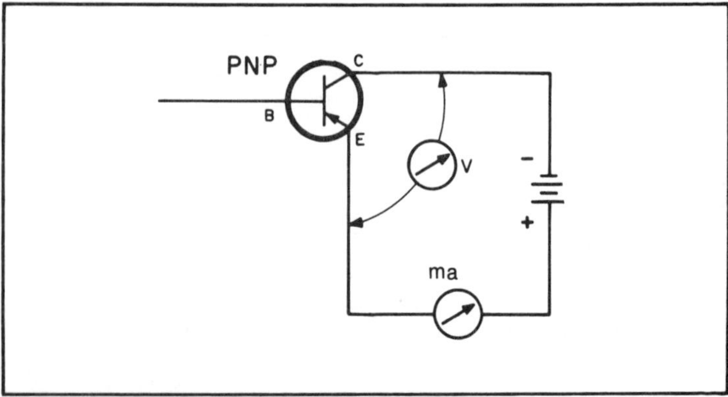

Fig. 4-21. Setup for measuring output resistance.

OUR TRANSISTOR MENU

One of the nicest parts about dining out is that you can get a menu and plan your own dinner. For your transistor menu we are going to serve transistor audio amplifiers in your next chapter. Following that, we are going to move back, stage by stage (in the succeeding chapters) until we have covered an entire receiver. And then, for your electronic dessert we will serve you with information on different transistor types and miscellaneous applications. We hope your appetite for knowledge will keep up with what we are going to put before you.

QUESTIONS

1. What is meant by fixed bias?

2. What is self bias? How does it differ from fixed bias?

3. Describe the circuit arrangement for determining the relationship between collector voltage and collector current.

4. What is a static test?

5. Explain the method for determining the load line for a transistor.

6. How can load lines be used to determine possible distortion?

7. What is the quiescent point?

8. What is a mutual characteristic?

9. What is stabilization? Why is it needed? What is a stabilizing resistor?

10. What is current feedback? What are its effects on a circuit?

11. In general, what is the relationship of input and output resistance in an amplifier circuit?

12. Transistors are current-operated devices. What does this statement mean?

Chapter 5

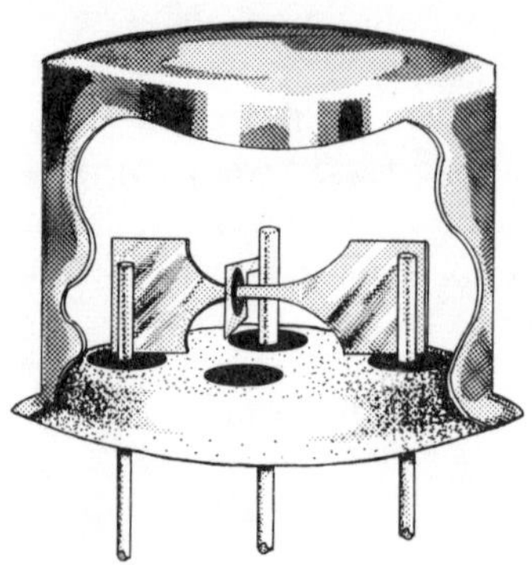

Audio Amplifiers

A TRANSISTOR IS TO A SIGNAL WHAT A LENS IS TO LIGHT—A MAGnifier. But what is this bit of magic whereby a few minerals are able to take a varying current and amplify it?

Let's start our investigation with Fig. 5-1. We're going to put a signal voltage right across R1. This varying voltage will modify the forward bias supplied by B1. The result of all this will be a varying current through the transistor and also through R2. Where, then, does amplification come in?

To understand this, there are a few items we must first consider. Our input circuit consists of the emitter and base of the transistor, battery B1 and resistor R1. The current that flows in this circuit is small—usually a matter of microamperes. How does this compare with the current in the output circuit? Here the current is much larger and is measured in milliamperes in this particular case. The output circuit consists of the emitter and collector, R2 and both batteries—B1 and B2.

This, then, is the first part of the answer to our question—why amplification? We have a much larger current in the output than in the input. But what more do we know about the currents? Both are varying at a rate determined by the input signal, for it is the input (ac) signal that modulates or changes the forward bias(dc).

Now examine the two resistors, R1 and R2 in Fig. 5-1. The resistor (R1) in the input circuit is much smaller than the resistor (R2) in the output circuit. We get signal voltages developed across

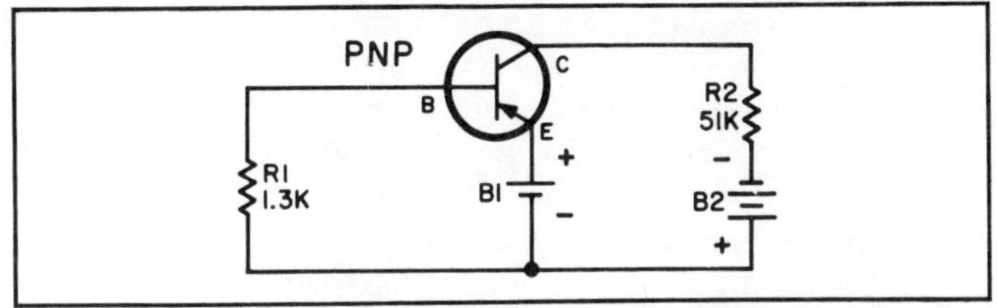

Fig. 5-1. This is an elementary amplifier circuit. An input voltage appearing across R1 will vary the forward bias. This will change the current through the transistor, developing a changing voltage across R2.

both of these resistors. What about the magnitude of these two voltages, both changing at the same rate? At the input, we have a small current going through a small value resistor compared with a large current going through a large resistor, at the output. And, since voltage is the product of current and resistance, we can see that the output voltage is going to be much larger than the input. In a representative transistor, the voltage across R2 will be several hundred times that across R1.

This reasoning on why a transistor amplifies may seem a bit too smug and too pat, so let's examine the problem somewhat further. To do so, suppose we ask a question: "Just what is the collector current?" Theoretically, if we had no base current, the current flowing to the collector in an npn transistor would be the emitter current. But we do lose some current, however small, to the base, and so the collector current is less than the emitter current by that amount. Offhand it might seem that, since the collector current is smaller than the emitter current, we could not possibly obtain amplification. The answer here lies in a comparison between the input and output resistances. We are not dealing with currents alone, but with voltages as well, since our currents are going to be made to flow through resistive or inductive components. Thus our varying voltage drop in the output is going to be much larger than that same varying voltage in the input simply because our output or load is much larger in value than the input resistance.

The fact that collector current is smaller than emitter current might seem contrary to what you may have learned in tube theory, but if you will think about it for a moment or two, you will see it is identical. Consider a pentode. Is the plate current equal to the cathode current? If there is no secondary emission to worry about and if the screen draws current, the plate current will be less than the cathode current but the tube doesn't know this and goes right on amplifying.

Does a transistor amplify? Not without some help. Not any more than a tube does. All that a transistor (or tube) can do is to control or vary the current supplied by a battery or power supply. A transistor or tube is just a link, permitting a small voltage (usually the signal) to control or manipulate or modulate the current furnished by a voltage source.

What is it that we have really accomplished? All we have done is to take a battery and a resistor and have managed to secure control of its current. You can invent an amplifying device of your own, if you wish. Figure 5-2 shows the basic problem. We have a resistor R1 connected across an ac source. This ac source could represent the signal supplied by a receiving antenna. A varying current will flow through R1. Nearby, we have R2 connected across a battery. Flowing through R2 is a direct current. Because R2 is much larger than R1 and because the current supplied by the battery is also much greater than that supplied by the signal, the voltage across R2 is several hundred times that across R1. Now all you have to do is to manage, in some way, to get the direct current through R2 to vary exactly in step with the current through R1 and you will have an amplifier.

AMPLIFIER ARRANGEMENTS

In all the circuits we have examined so far we have used one special type. You may have noticed that in each case we always specified one element—the emitter—as common to both the input and output circuits. For this reason it is called the common-emitter circuit. Since ground is often a widely used reference point, the circuit is sometimes referred to as a grounded emitter.

Because the transistor has three elements—emitter, base and collector—we can use any of the three as the common element.

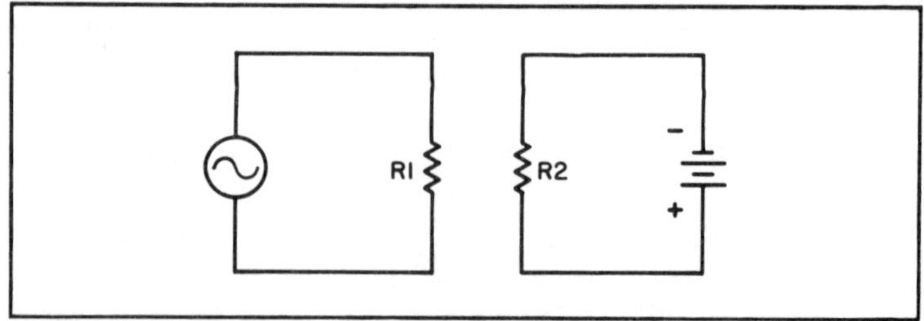

Fig. 5-2. Want to invent an amplifying device such as a tube or transistor? Solve the basic problem shown here and you are on your way to fame and fortune. All you need do is to make the varying current through R1 control the current through R2.

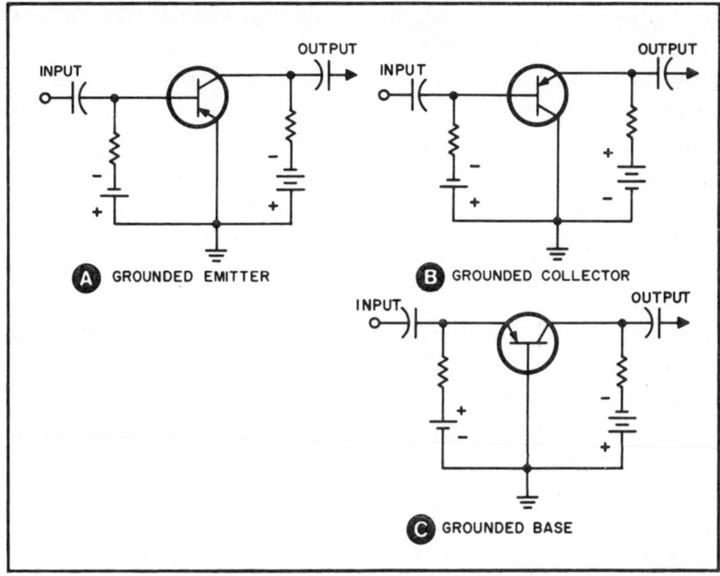

Fig. 5-3. Basic transistor amplifier arrangements.

Figure 5-3 shows the three possible circuit arrangements. Each of the circuits has its own characteristics. However, the common or grounded emitter circuit is the most widely used.

The three basic circuits have certain similarities. The common emitter and common collector arrangements are identical except for the transposition of the output elements. Thus, you could easily change a common emitter circuit into the form we call common-collector just by transposing a few leads. Similarly, the common emitter and common base could be changed, one to the other.

PHASE

The word phase is used so often in electronics that we must understand just what it is trying to tell us. Phase is used to describe the relationship existing between a pair of currents, a pair of voltages, or a current and a voltage. Sometimes a voltage is compared to what it was when it started or what it will be at some time other than this moment.

Now that we have managed to confuse you, let's unravel this tangled web of words. The classic description of phase is to compare it with a seesaw. If you sit on one end and we on the other there is no possible way in which we can both rise or fall at the same time. When you go up, we go down. The two ends of the seesaw are in opposite phase, or out of phase—call it what you like. If I climb up when you march down—we are out of phase. If you save while we spend—we are out of phase. If you charge a battery while we discharge a battery—the battery voltages will be out of phase. One will be increasing, the other decreasing.

We can also have in-phase conditions too. If we both go up in an elevator, we are in phase. We are rising at the same time. If we applaud a show at exactly the same time, we are in phase.

Suppose we have two ac voltages of identical frequency, and let us imagine that they both start from zero at the same time. If they both move in a positive direction, they are in phase. If one voltage becomes more positive while the other becomes more negative, they are out of phase.

Examine the common-emitter circuit shown in Fig. 5-4. When a varying signal appears across R1, we will have a varying signal across R2 as well. Suppose point A (resistor R1) becomes more positive with respect to point B. When this happens, point C (resistor R2) will become more negative with respect to point D. But we have the input voltage across R1 and the output voltage across R2. Thus, the two voltages are out of step or out of phase. But since the voltage across R2 is a magnified replica of that across R1, we can say that the common-emitter circuit in Fig. 5-4 has given us not only amplification, but phase inversion as well.

Now suppose we were to take the common-emitter circuit of Fig. 5-4 and just transpose the leads to the emitter and base. We would then have a common-base circuit. But what happens to the current through R1 when we do that? It just flows in the opposite direction. As a result we have changed the polarity of the voltage across R1 so that it is now in step or in phase with the voltage across R2. We can understand this a bit more easily if you will remember that in going from the common-emitter to the common-base circuit, we made no change in the collector or its load resistor or in the output battery.

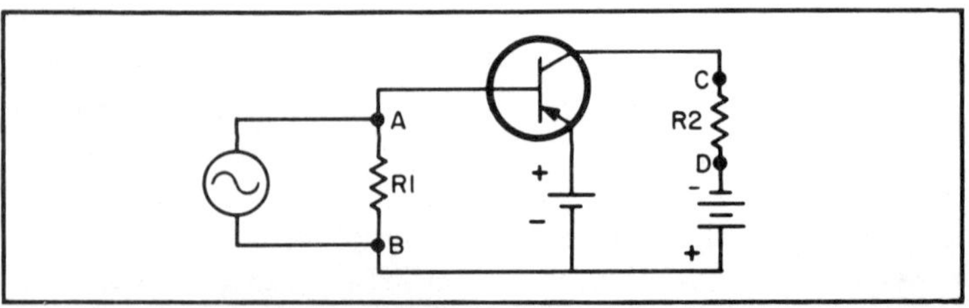

Fig. 5-4. A slight case of phase inversion. When point A becomes more positive [is positive-going], point C will become more negative [is negative-going].

Now what about the common-collector circuit? It has no phase reversal either. And the reasoning is exactly the same as that used for the common-base arrangement. To get a common collector, we take the common emitter and transpose emitter and collector leads. The effect of this is to transpose the collector load resistor. But if we do this, we also transpose the polarity of the voltage across the collector load. This puts it in phase with the input voltage.

THE COMPLEMENTARY SYMMETRY PRINCIPLE

The three basic amplifier arrangements—common emitter, common base, and common collector—were shown in Fig. 5-3 as pnp units. We could have used npn transistors just as well, once more keeping in mind that npns and pnps may require the same voltages but opposite polarities. And because the polarities are transposed, so are the currents. We might even say that the two types are mirror images or complementary. We are going to build circuits around this knowledge. The idea of complementary units is unique to transistors since we have nothing like it in vacuum tubes. Tubes require a positive voltage on the plate—transistors can have either positive or negative voltage on the collector depending on whether we use a pnp or npn type.

BIASING THE AMPLIFIER

In radio tube circuits we will sometimes come across a tube designed to operate with zero bias. This is a condition you will not find with transistor audio amplifiers since bias is an essential part of their operating conditions. Let us see why this is so.

In Fig. 5-5 we have a circuit that is normal in every respect, except that we have no battery for forward bias. If the input signal is symmetrical (such as a sine wave) then, during half the input cycle the base-emitter circuit will be forward-biased, but during the other half of the cycle it will be reverse-biased. During forward biasing we will get current flow through R_L, the collector load resistor. But during reverse biasing of the input by the signal, only an extremely tiny current will flow through the collector load. Since the output signal does not resemble the input, we have a severe case of distortion. Actually, the circuit in Fig. 5-5 behaves more like a rectifier, although there is amplification for part of the cycle. (In Chapter 6 this action is used for detection.)

What should the relationship be between forward bias and signal voltage? Since the input signal alternately adds to and subtracts from the forward bias, we must make sure that the bias

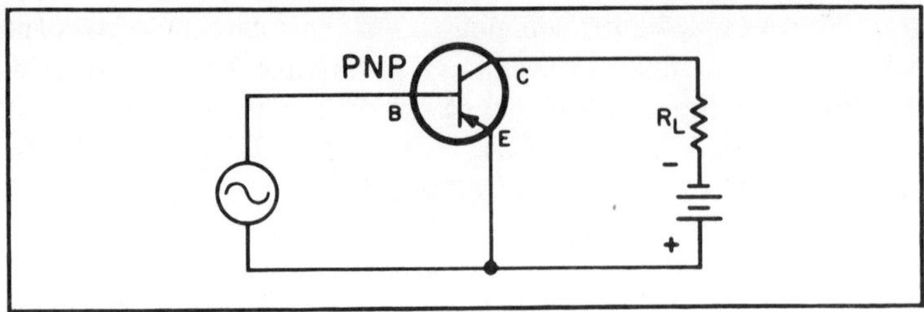

Fig. 5-5. The input circuit is not properly biased since the ac input will supply, alternately, forward and reverse bias.

voltage is always large enough to give the signal something to work with. Since you may be snowed under by this torrent of words, let's consider several possible situations. Suppose the bias is 2 volts and we have an input signal of 1 volt, peak-to-peak. Assuming the same sort of waveform we had earlier, the net result of combining the ac signal plus the dc bias would mean (as shown in Fig. 5-6) that half the time our forward bias would be:

$$2 + .5 = 2.5 \text{ volts}$$
$$2 - .5 = 1.5 \text{ volts}$$

Our forward bias now ranges between 1.5 and 2.5 volts. Because our bias battery is larger than the input signal voltage, we always meet the condition of having forward bias in the circuit.

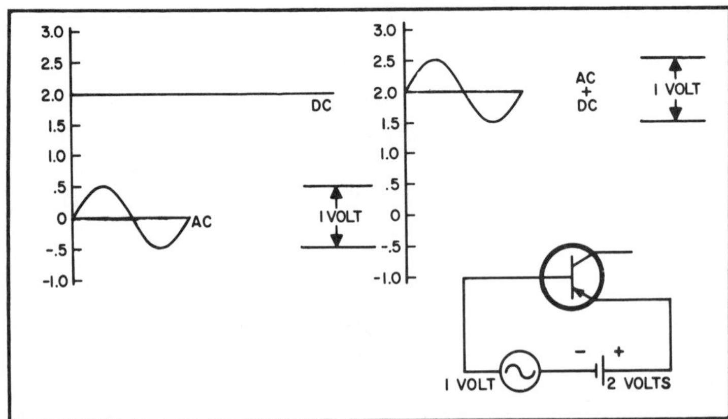

Fig. 5-6. For proper biasing, the signal voltage must be less than the forward biasing voltage. Note that the dc bias is far above the zero reference line. The signal voltage varies above and below the reference. In the graph you can see that the combination of dc bias and ac signal does not permit the resultant voltage to drop below zero.

But what if the opposite took place? What if we had a bias of just 1 volt dc, but a signal voltage of 3 volts, peak-to-peak. What has happened in our input circuit? We now have:

$$1 + 1.5 = 2.5 \text{ volts}$$
$$1 - 1.5 = -0.5 \text{ volt}$$

Figure 5-7 shows graphically, the result of such operation. So, we always want to make sure that our bias will be larger than the highest peak-to-peak signal voltage we are going to have.

Now that we have convinced you that an improperly biased transistor results in a distorted or a rectified output signal, it is our sad duty to inform you that what appears to be a vice can, if we so wish, be virtue triumphant. And, if you're about to leap forward to the conclusion that what we have in mind is some sort of rectifier or detector, control your reflexes, for this is a chapter on audio amplifiers. But, while you are still tantalized about this forthcoming tidbit of information, let's take a small sidepath. Don't fret about this detour, since it is really just a shortcut to the main highway.

CLASSES OF AMPLIFIERS

There are many ways of classifying audio amplifiers. We can talk about voltage and power amplifiers, driver and output amp-

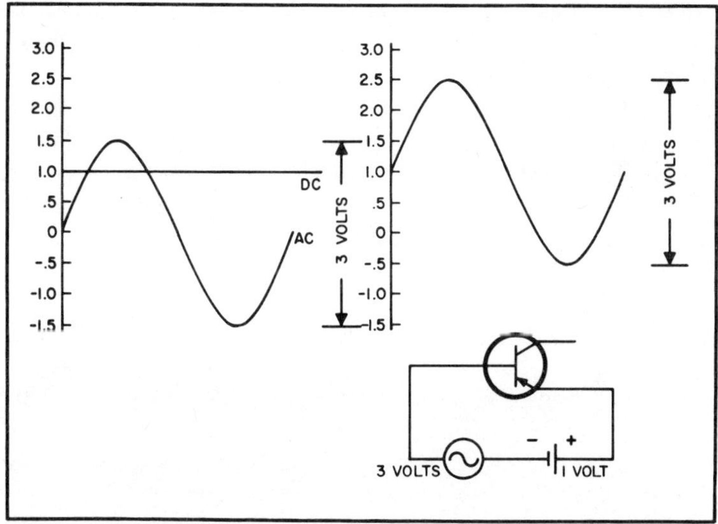

Fig. 5-7. When the signal is larger than the bias voltage, the transistor input will be both forward and reverse biased.

lifiers, push-pull and single-ended. One other way is to arrange them in accordance with their bias requirements. For transistors, where we seldom stray far from the word "bias," this is a very attractive thought indeed. And, with a strong penchant for original thinking, we reach right over into vacuum-tube theory, and borrow a classification, lock, stock, and element.

A favorite vacuum-tube type of classification is to call amplifiers class A, B, or C. This doesn't mean that A is better than B, or that B is slightly superior to C. In vacuum-tube theory, a class A amplifier is one so biased that plate current flows no matter which part of the input cycle is tickling the grid at the moment. Thus, whether the input signal is crossing its zero axis, is at maximum positive or maximum negative, we always get more or less current flowing to the plate. A class-B amplifier has plate current flowing for only fifty percent of the input cycle, while for class C, plate current exists for only a small fraction of the input cycle.

But how do we arrive at that happy condition in which we can make the plate current do our bidding? How can we take a single tube and make it work as class A, or class B, or class C, just as we wish? The clue to this—no, the whole answer to this—lies in the bias. Very high bias and you will probably have class C. Very low bias, class A. And class B somewhere in between.

If, from your studies of vacuum-tube theory you know all this, be of good cheer for you to also know this information for transistors, and without effect on your part. In tubes, we talk about bias controlling plate current. In transistors, it is the amount of forward bias that determines collector current.

In Fig. 5-8 we have a drawing which demonstrates this for you graphically. What is the class of operation? All you need do is to select the correct amount of forward bias for the input circuit. However, we have been a little crafty about this since we haven't really told you the whole story. We start with what we fondly believe to be the correct amount of bias for the class of operation we want, and then (this is the sneaky part) we make sure that our input signal is never large enough to distort the output waveform. Take the case of class-A operation in Fig. 5-8. We very carefully chose an amount of forward bias that put us as close to the center of the collector current curve as possible.

Now work along with us with the graph in Fig. 5-8. Locate the dashed line marked "base bias line." What is this line and just what does it represent? This line is our forward bias. If you will go back to Fig. 5-6 you will see that we have a straight line marked "dc." We

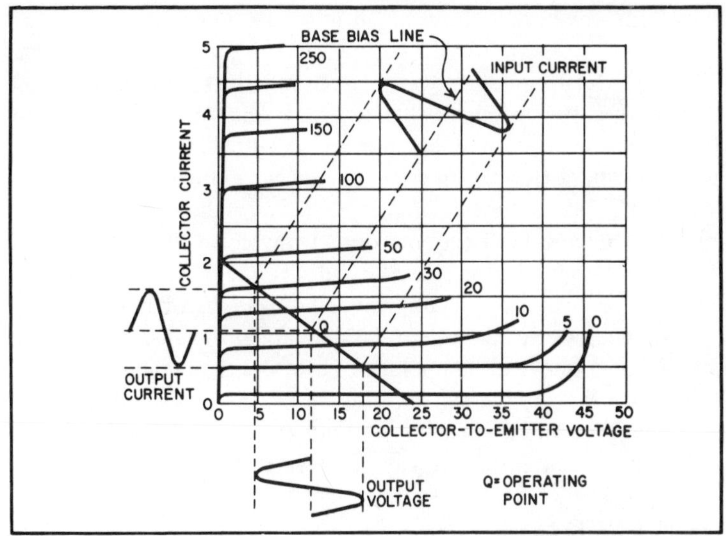

Fig. 5-8. The class of amplifier operation is set by the operating point Q on the load line. Changing the base bias will move this point, resulting in a different class of operation.

have a similar line in Fig. 5-7. This is the type of line we are using in Fig. 5-8.

Now suppose that to the dc represented by the base bias line we add an ac voltage. The result will be a variation around the base bias. To get the appearance of the output current and the output voltage, we can project to the left and also downward. If you will look at the dashed lines in the drawing, you will see that they center equally around the base bias line. Also note that we are working along the straight portion of the collector current curves. As a result, our output waveform is a rather decent replica of our input waveform.

But what if we change our forward bias? We can either increase the bias or decrease it from the condition shown in Fig. 5-8. As a start, let us say that we increase the forward bias. The effect of this will be to move the base bias line down the slope of the load line. To see what will happen, just imagine the letter Q (our operating point) sliding down the load line and coming to rest somewhere near the bottom. We could still project our input current waveform onto the load line, but a projection of the output current and output voltage waveforms would show severe distortion. Similarly, if we decreased the forward bias, the base bias line would move up on the slope of the load line, once again producing a distorted output waveform—even though the input might be undistorted.

Now here is where the subtle part comes in. Note that in Fig. 5-8 our modulating wave never takes us into the saturation or cutoff regions of collector current. How do we know this? Look at the vertical dashed lines moving up from the output voltage waveform and note where they touch the collector current curve. These two lines come close to the beginning and the end of the curve. But if our input signal were too large, our output waveform would be clipped top and bottom. And, since the output would no longer be an enlarged replica of the input, we would have distortion.

How do we get the other classes of operation—classes other than A? Simply move the operating point Q up or down on the load line. Class B works at the cutoff point—class C beyond that. We also have one more class—AB—with an operating point on the collector current curve halfway between that of A and B.

CLASS A VS CLASS B

There is just one class of operation that does not result in distortion. This is class A. All the others—AB, B and C—give us an output waveform that is an amplified, but amputated version, of the input. But before we rejoice about class A, before we decide that that's for us, consider its limitation. Make the input signal too strong and we will get an output waveform that will be clipped top and bottom—a distorted wave. This is the disadvantage. With class A we actually cut the useful length of our collector current curve in half. We start our bias at its center point. It's just as though we had a street in which to practice running, but, instead of starting at one end and running the full length, we started at the center, with the option of running to either end.

How can we get around this seemingly dead-end situation? We can run our audio amplifier class B. But since class B gives us what is tantamount to rectifier operation, one transistor with half a waveform output won't be enough. We get right past this problem just as neatly as you please by using two transistors in class B push-pull. One transistor supplies the lower half of the output wave; the other supplies the upper half. But please don't get the idea that just class B works in push-pull. We have every class, A, AB, B, and C using push-pull. Unfortunately, class C produces so much distortion that we can't use it for audio work, but it does nicely for rf.

VOLTAGE AMPLIFIERS AND POWER AMPLIFIERS

Practically every circuit in a transistor radio receiver is an amplifier, starting at the antenna and going right on out to the speaker. And, if the detector is a transistor, we can make the statement a unanimous one.

In any radio set, the objective is quite simple. The problem is to take the signal off the antenna or loopstick and amplify it—and to keep right on amplifying it. Granted that various changes do take place. We get frequency conversion up at the front end and signal rectification somewhere along the middle of the chain of events. But as we move, stage by stage, in the direction of the speaker, we amplify the signal. The reason for this is that we must have a signal strong enough to dominate the activities of the output transistor.

With this idea in mind, we can roughly classify amplifiers as voltage or power types. Note the word "roughly," used deliberately, for this separation of amplifiers into voltage and power types is a crude designation, but a handy one, nevertheless. (In the case of transistors, it might be more accurate to call them current, rather than voltage amplifiers.) In the sense, though, that our signal is a voltage, and that our transistors are intended to augment this voltage, we have voltage amplifiers. And, because our audio output transistor must furnish the energy to move the voice coil, it is a power amplifier. The separation between voltage and power amplifiers is not sharp and clear-cut. Voltage amplifiers supply some power gain and our power amplifiers do deliver some voltage gain.

There are many ways in which we can classify amplifiers. We can pigeon-hole them quite neatly by bias (class A, B, etc.) or by whether they are primarily voltage or power amplifiers. We can describe them as single-ended or push-pull. A single-ended amplifier uses a lone transistor to deliver power to the speaker. Push-pull uses a pair. Or we can talk about amplifiers in terms of their functions—an output amplifier or a driver amplifier. And, just as some people need four or five names to achieve complete identity, so too do we sometimes take all of these designations, and come up with a mouthful such as: class A, single-ended, power amplifier.

CLASS A, SINGLE-ENDED, POWER AMPLIFIER

This may seem like a rather elaborate name to hang on the inoffensive little circuit shown in Fig. 5-9, but we can extract a lot of information from that name. Class A gives us a clue about the forward bias and our location on the collector currrent characteristic. Single-ended tells us not to look for more than one transistor.

Power amplifier is our clue to look for a speaker or output transformer as the load.

Let's examine the circuit of Fig. 5-9 to get acquainted just a little better. Our component parts are simple enough and few enough. A small assortment of resistors, capacitors, a transistor, transformer, and a battery, and that is all there is to it. Starting with the capacitors we see that we have two of them. Both are electrolytics. C1 is a coupling capacitor from the collector of the preceding (driver) stage. C2 is the emitter bypass. With electrolytics we always have the problem of polarity. We've marked C1 and C2 carefully, but the question still remains as to why they are positioned in this way.

Let's start with C1. One side of the capacitor connects to the collector of the pnp driver. But this collector is wired to the negative terminal of the battery. The plus side of C1 is soldered to the junction of R1 and R2. This is a plus point with respect to the top end of R1.

Now what about C2? Current flows through R3 from the emitter toward ground. Because of this direction of current flow, the top end of R3 is minus, the bottom end plus.

If you will look carefully, you will see that R1 and R2 are in series, but that this series combination is connected directly across the 4½-volt battery. It is true that the base is tied right onto a plus point on the voltage divider, but because of the voltage drop across R2, the base is less positive than ground. Ground is the maximum plus point since it is attached to the positive terminal of the battery.

How does the transistor get its forward bias? In two ways. Our

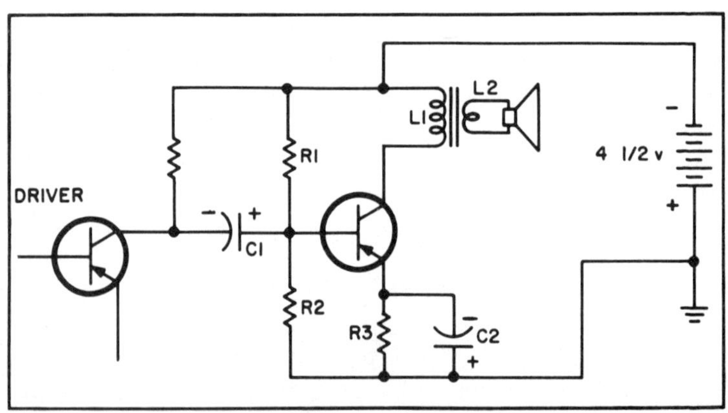

Fig. 5-9. Single-ended class-A power amplifier R-C-coupled to the preceding stage.

voltage divider, R1 and R2, supplies fixed bias. R3, the emitter resistor, takes care of self bias. The values of R1 and R2 are chosen so as to put the operating point of the transistor at the center point of the collector current curve (with the help of R3). This gives us our class-A operation.

The collector is transformer coupled to the speaker through an output transformer. The transformer is a stepdown type. L1 has a primary impedance of several hundred ohms. L2 is just a few ohms.

What other information can we get out of Fig. 5-9? The transistor is a pnp type. If we were to use an npn we would still work with the same components, but rearranged. We would need to transpose the two electrolytics and the battery. We probably would need to alter the values of R1 and R2 to meet the bias conditions of the new transistor.

RESISTANCE COUPLING AND TRANSFORMER COUPLING

The circuit of Fig. 5-9 is resistance-capacitance (RC) coupled to the preceding driver stage, but a transformer could have served just as nicely. We see this new arrangement in Fig. 5-10. As you see, it is almost the same as the RC stage. We've removed coupling capacitor C1 and put in transformer T1.

The change is an interesting one. T1 is a stepdown type since the collector of the preceding driver stage is a much higher impedance circuit than the base-emitter circuit of the power amplifier. As before, R1 and R2 form a voltage divider shunted directly across the

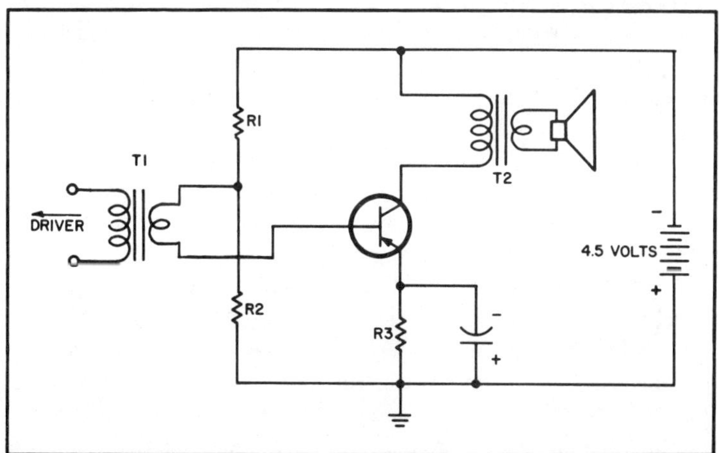

Fig. 5-10. Single-ended class-A power amplifier transformer-coupled to the preceding stage.

battery. In series between the dc voltage supplied at the junction of R1 and R2 we now have the ac signal voltage delivered by the secondary of T1. This is no great development since we know that the purpose of the signal is to modulate or change the forward bias.

In vacuum-tube receivers, RC coupling to the output stage is so customary that an occasional transformer coupled job causes some eyebrow lifting. In transistor receivers, though, transformer coupling is common.

STABILITY OF OPERATION

You would think that having gone to the trouble of determining the correct amount of forward bias to put on our transistor (thereby establishing the operating point on the collector current characteristic) this operating point would remain as fixed as a dab of glue. The transistor, though, is temperature sensitive. This means that as temperature goes up, so does collector current. Back in Chapter 4, we told you about feedback techniques for stabilizing the transistor. But there are other methods. One technique is to use a thermistor. This is a semiconductor which functions as a temperature-sensitive resistor having a negative temperature coefficient. Translated, this means a thermistor is a component whose resistance goes down when the temperature goes up, and vice versa.

HOW MUCH CHANGE?

This doesn't tell the whole story. Metals have a positive temperature coefficient, meaning, of course, that their resistance tags along after temperature, rising when temperature goes up, decreasing when temperature goes down. But the amount of resistance change is small. Thermistors, however, have a large temperature coefficient, just another way of saying that they are very sensitive to temperature changes. To give you an example, in the range from zero to 200 degrees, a platinum wire would have an increase in resistance amounting to a fraction of an ohm. A similar volume of thermistor material, for the same temperature increase, would change from about 8,000 ohms down to about 11 ohms.

Thermistors are made in the shape of discs, washers, and rods, of manganese, nickel and cobalt oxides. As in the case of resistors, suitable binder material is added, and then the combination of chemical and binder is extruded, pressed or molded into the desired shape.

Now let's see how the thermistor does its job. In Fig. 5-11 we have a thermistor, quite properly shown as a variable resistor, shunted across R2. R1 and R2, as we know, supply forward bias to the base-emitter input circuit. And, because we have a pnp transistor, we want our emitter to be positive with respect to the base. The bottom end of R2 is our most positive point since it is connected directly to the plus terminal of the battery. As we move up on R2 we become less positive, since, if we continue long enough, we will go right on through R1 to the negative terminal of the battery.

Collector current, though, depends upon the proper forward biasing of the base-emitter circuit. If the forward bias should decrease, so would the collector current. The thermistor in Fig. 5-11 is quite able to go from a value of almost zero ohms to a value much higher than that of R2. As long as collector current is normal (by normal we mean the correct amount of collector current), the resistance of the thermistor will be so high that its paralleling effect on R2 will be negligible. Or, if not, then the combination value of thermistor and R2 can be selected so that the two in parallel supply the correct amount of resistance.

Now suppose that an increase in temperature causes the collector current to increase. This same temperature rise decreases the resistance of the thermistor—fast and substantially. Because of this, the total resistance between base and ground (R2 in parallel with the thermistor) becomes smaller. But this has the effect of lowering the amount of forward bias since the base now approaches ground. But ground is positive and the base needs to be negative.

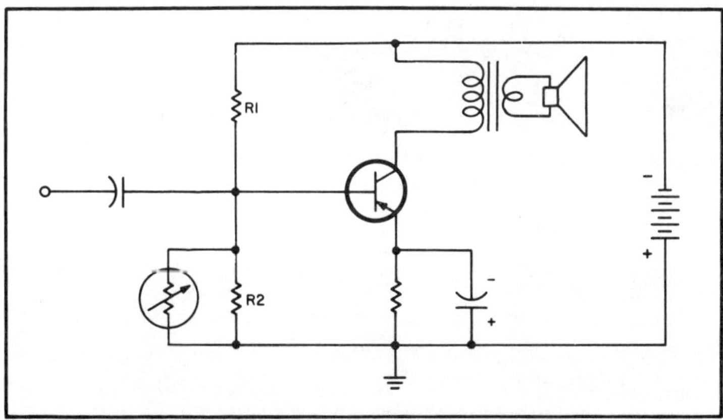

Fig. 5-11. Thermistor shunted across R2 prevents thermal runaway. Electrically, the thermistor is in parallel with R2. Physically, it is placed near the power transistor and is thus immediately affected by any temperature rise of this component.

Fig. 5-12. The purpose of the heatsink is to dispose of heat as quickly and as effectively as possible. Heatsinks are often made in a corrugated or ribbed style to provide as much surface area as possible. A metal chassis is also used as a heat sink (Delco Radio Div., General Motors)

The reduction of forward bias cuts the collector current down to a smaller value.

HEATSINK

Since what we have said has given you the impression that heat and transistors had best be kept apart, it may not be surprising to learn that output transistors can get hot and do get hot. Power transistors, though, often have collector current ratings measured in amperes, and where we have amperes, plus resistance, we have heat. To dissipate the heat we do something that heating engineers and automobile engineers have been doing for a long time. We use a large metal surface to radiate the heat, and the more surface area, the better. Sometimes the chassis is used to conduct heat away from the transistor. In some cases the power transistor is mounted on what seems to be a series of metal fins. A unit of this sort is shown in Fig. 5-12.

EFFICIENCY—CLASS A

The power to operate a transistor radio comes from its battery. The ac power or the signal power used to drive the voice coil of the speaker is derived from that battery. Basically, what happens in a transistor radio is that the power supplied by the battery is modified or altered by the signal, in cooperation with the transistors. But here, as in everything else, we must pay a price. Our battery delivers dc watts. Our speaker needs ac watts. In the conversion, the transistor will exact its little toll. The ratio of these two values, the output power divided by the input power, is always less than one. In the case of the single-ended class A power amplifier, the maximum efficiency is 0.5 or one half. Thus, if our battery supplies 6 dc watts , and our audio power at the output is 3 watts, then 3/6 equals 0.5. We convert this to a percentage by multiplying our answer by 100. The maximum efficiency of our single-ended class-A power amplifier is 50 percent, but usually less. And if you think this is a low value, some steam locomotives and machines clock in at 12 percent efficiencies. However, if efficiency is a consideration (and there are others) we can move on to push-pull amplifiers.

THE PUSH-PULL AMPLIFIER

Our single-ended amplifier works class A because this is the only way in which it can work. We can't operate it class AB, B or C without running into some waveform distortion. However, this limitation does not cause trouble in some rf amplifiers.

The circuit of a class-B power amplifier is shown in Fig. 5-13. We've drawn this circuit for you in a rather neat way, but unfortunately its very neatness hides the fact that the push-pull amplifier is nothing more than our single-ended job multiplied by two. To see that this is really so, we have separated the biasing system into two sections, as shown in Fig. 5-14. Since we have pnp units, we want our emitter to be maximum positive, the collector maximum negative, and the base somewhere in between. But isn't this the same arrangement we had for the single-ended amplifier? We've omitted the emitter resistor, but all you need do is to break open the line

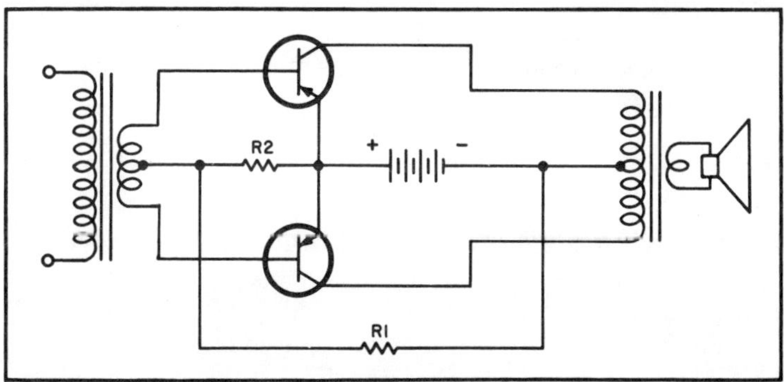

Fig. 5-13. Pushpull amplifier. Class of operation is determined by the values of R1 and R2. Because of the way the circuit is drawn, it is difficult to see that R1 and R2 act as a voltage divider across the battery, to supply forward bias to the two transistors.

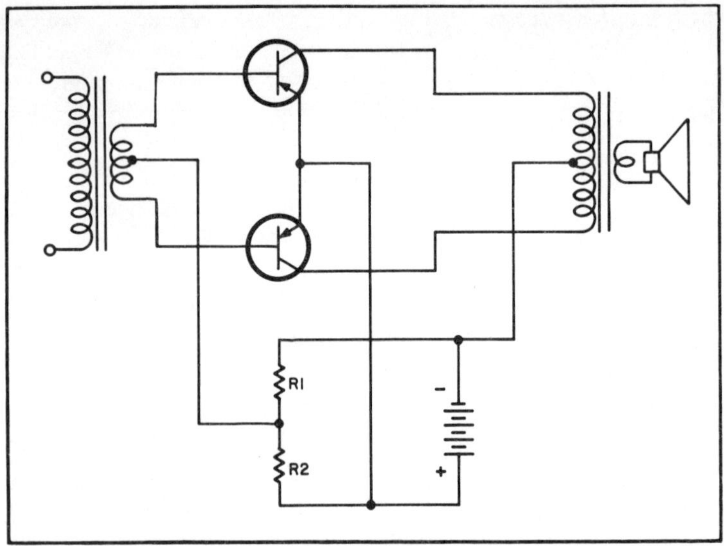

Fig. 5-14. The circuit of Fig. 5-13 has been redrawn to show the voltage-divider action of R1 and R2. The emitter is maximum positive. The voltage divider makes the base less positive than the emitter. Collector is maximum negative.

between the common emitter connection and the plus terminal of the battery and insert it. And, if we need to or want to, we can also add a thermistor.

We haven't told you what class of operation we've been using in the push-pull circuit of Fig. 5-14. Actually, you can't tell without a score card. The class will depend on the relationship of R1 to R2. To understand this, imagine R1 and R2 as a single resistor (but with the value of R1 and R2 in series) and the fixed connecting point as a sliding arm on that single resistor. As we move the slide down toward the bottom end (the end connected to the plus terminal of the battery) we keep reducing the forward bias. This drives the transistor collector current down toward the cutoff point. If we move the slide arm up toward the top end of the resistor, we increase the forward bias, so that the collector current goes to saturation. But what are we really doing? We've actually been sliding up the operating point Q (see Fig. 5-8 once again) from its point at cutoff to its point at saturation.

But if we can determine our operating point in this way, what we are really doing is establishing the class of operation we will use. Incidentally—we didn't have to go as far as we did. If we started at cutoff, and moved up to the center point of the collector current characteristic, that would have been far enough.

EFFICIENCY—CLASS B AND AB

Class B efficiencies run up to a little more than 75 percent while class AB comes somewhere between class A and class B. Class C is the most efficient, but as we told you earlier, it isn't used for audio work, because of the distortion it produces at audio frequencies.

In class B, when one transistor is hovering around the cutoff point, the other transistor is doing an honest day's work. No criticism here, though, since each transistor takes its turn in delivering collector current to the primary of the output transformer.

Aside from the greater efficiency of class B push-pull contrasted with class A single-ended, the push-pull arrangement has other attractive features. The core of the output transformer for push-pull can be smaller than the core of a similar transformer for single-ended. The reason for this lies in the effect direct current has on the core of a transformer. All currents, whether ac or dc, carry a magnetic field along with them. But in a transformer, only a varying magnetic field (such as that supplied by ac) is of any use in the transfer of energy from primary to secondary. The steady magnetic field of dc not only does not transfer energy, but it actually preempts some of the iron core for itself. Thus, if we have both ac and dc flowing through the primary of a transformer, we can get a greater transfer of energy if we simply remove the dc. In push-pull, we have two direct currents flowing, one from each collector (or toward each collector). But these two currents have magnetic fields which oppose each other. This leaves the core free to take care of the ac component—that is, the varying current due to the signal.

POWER OUTPUT

The power we can get out of a transistor depends on its class of operation, but this isn't the whole story. Power output is determined by the amount of heat the transistor can dissipate. In turn, this depends on the cooperation we give the transistor in the way of a heatsink. And, to compound the confusion, not all heatsinks have the same thermal efficiency. In other words, some heatsinks get rid of heat better than others. If the chassis is used as the heatsink a heat-conducting insulating washer is inserted between the chassis and the power transistor. The washer is made of mica, anodized aluminum, Teflon, Mylar or fiberglass. To improve heat conductivity, each side of the washer should be coated with a thin film of silicone lubricant. If we didn't use a lubricant, the space between the base of the transistor and the chassis might become a dead-air region, and an air pocket of this sort could effectively block the transfer of heat to the chassis. When you replace power transistors, coat the washer with silicone on both sides.

Unlike transistors used in stages preceding the audio output, power transistors usually have but two leads—base and emitter. The collector is connected to the case of the transistor to help conduct heat away from the junction. There are some power transistors made with three leads, but these are not as common as the two-lead variety. Sockets can be used with power transistors but, when currents are measured in amperes, it doesn't take too much contact resistance to produce an unwanted (and, frequently, unexpected) voltage drop. In some cases it may be more practical to weld or to solder connections directly to the leads of the power transistor or to use machine screws and lugs.

OTHER PUSH-PULL ARRANGEMENTS

The push-pull circuits we've examined are common-emitter types. Common-base and common-collector arrangements could be used, but you will find the common emitter most often.

As in the case of tubes, transformers represent a natural application in transistor push-pull circuits. Transformers are not only highly convenient for impedance matching but also supply phase inversion. But, no sooner do we get a component to do a good job, than all sorts of schemes get underway for eliminating that component. Transistors lend themselves very nicely to this bit of skullduggery if we take advantage of the fact that in npn and pnp transistors, currents move in opposite directions.

The pushpull circuit, known as complementary symmetry, is shown in Fig. 5-15. A careful look shows that the two transistors are really connected in parallel.

To understand how this circuit works, consider an input signal. Depending on its polarity at the moment, it will either increase or decrease the forward bias. But if that signal is applied to the base-emitter input of two different transistor types, such as an npn and a pnp, it will increase the forward bias of one and decrease the forward bias of the other during one-half of the cycle and produce exactly the opposite effect during the other half of the cycle. But this is exactly what a transformer will do for us.

In the circuit of Fig. 5-15, our output is taken from the common emitter, joined by the primary of the output transformer. If you will trace R1 and R2 (for both transistors) you will see that once again we have our usual voltage dividers, R1 and R2 for the pnp being

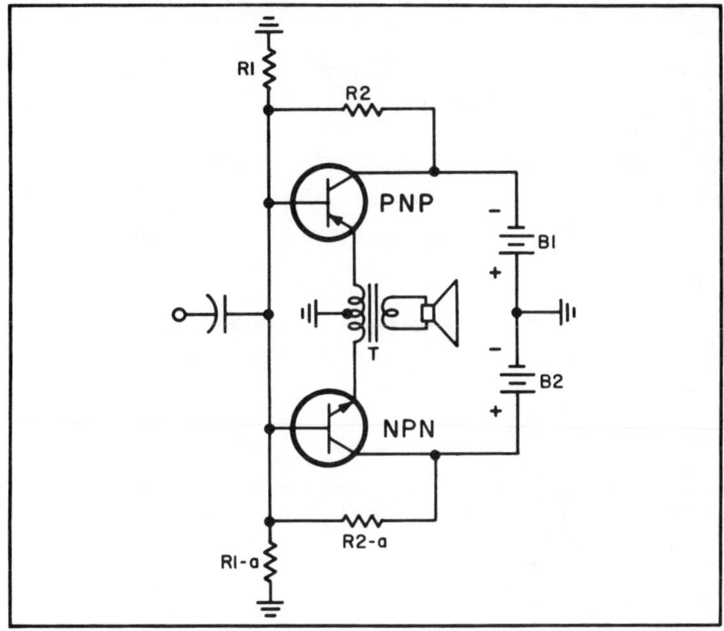

Fig. 5-15. Complementary symmetry arrangement takes advantage of opposite current flows in npn and pnp transistors, manages to eliminate the input pushpull transformer.

shunted across B1, while R1-a and R2-a for the npn are in parallel with B2.

ADJUSTING THE BIAS

It doesn't take much of a change in forward bias to push us up or down on the load line. And, even using the bias voltage recommended by the manufacturer doesn't always ensure that we will get the class of operation we want. We can get out of this difficulty by making one of the biasing voltage divider resistors variable. Unfortunately, this makes biasing rather critical.

A better setup is shown in Fig. 5-16. Here we have our voltage divider, R1 and 2, furnishing forward bias to both pnp transistors. In series with R1 and R2, we have a third resistor R3. The value of this resistor can range from practically zero, to almost full value, depending on the setting of the arm of the bias adjustment potentiometer. Note that the bias adjust pot is shunted across R3. With one position of the arm of the bias adjust pot, R3 is shorted. At the opposite position, R3 is almost full value, since the resistance of the bias adjust pot is so much greater than R3.

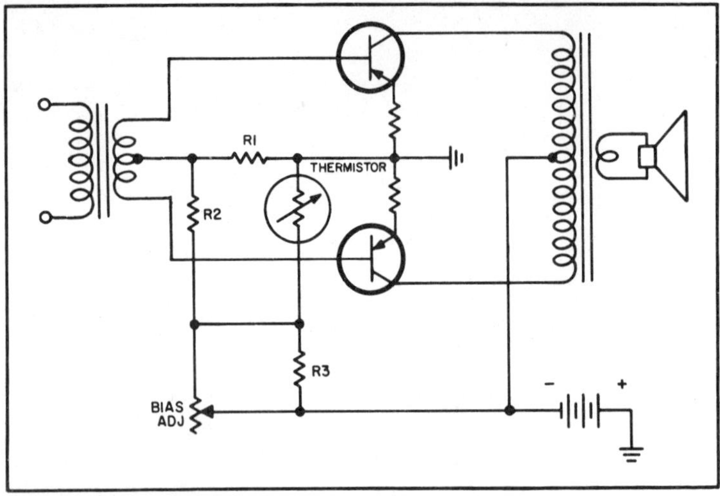

Fig. 5-16. Bias-adjust potentiometer permits small change in the forward bias of the push-pull transistors.

THE DRIVER STAGE

Feeding the power output, whether this is single-ended or push-pull, is an audio amplifier known as a driver. Driver amplifiers are invariably class A since a single transistor is used. A driver circuit is shown in Fig. 5-17. The input is resistance-capacitance coupled to the preceding stage. In a transistor receiver this could be the detector. R1 and R2 form the usual voltage divider to supply forward bias. The interstage transformer, T, can be used to drive a pair of push-pull transistors.

Circuitwise, how new or different is the driver compared to

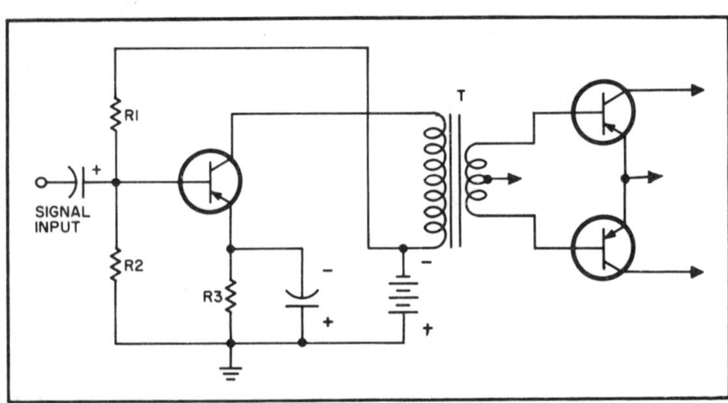

Fig. 5-17. Audio driver stage.

single-ended power output? Compare the two circuits, Figs. 5-9 and 5-17 and you will see the remarkable resemblance.

CONTROL POSITIONS

There are three types of controls you will find associated with audio amplifiers. One of these, the bias control, has already been mentioned, and is found in connection with power amplifiers. The other two, tone and volume (or gain) controls, are part of driver circuitry or output stages.

Two tone control circuits are shown in Fig. 5-18. They both work in exactly the same way. They supply bass and treble boost, although actually, neither of the two circuits does anything of the sort. Both work because they bypass higher audio frequencies to ground, the amount of "highs" being shunted in this way depending on the position of the potentiometer in Fig. 5-18A, or the setting of the rotary switch in Fig. 5-18B. As more highs are bypassed, there is a seeming boost of the lower frequencies. The tone control

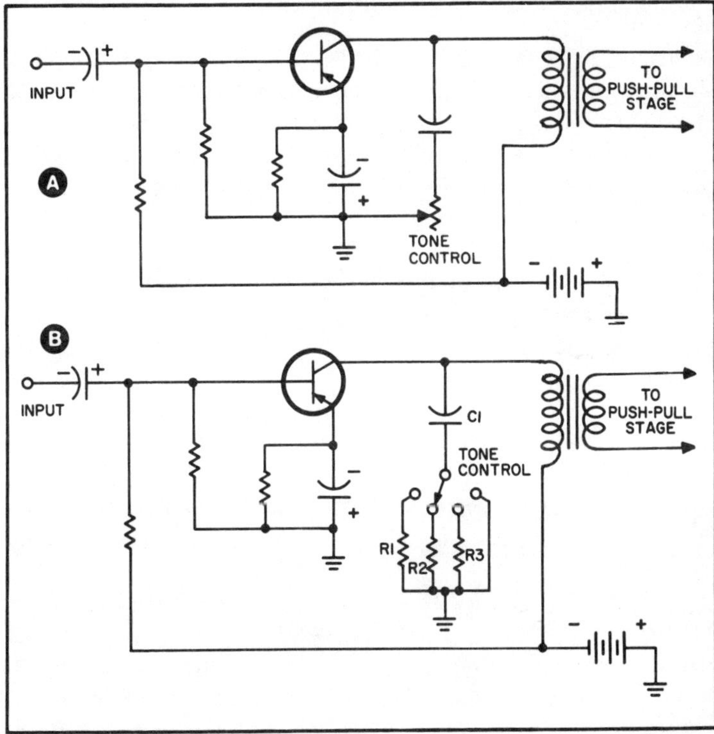

Fig. 5-18. Two types of tone control circuits.

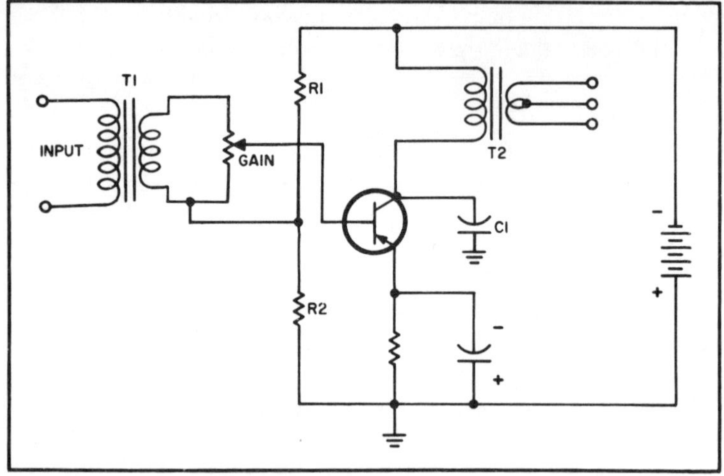

Fig. 5-19. One possible position for the gain control is at the input to the driver stage.

consists of a capacitor in series with a resistor. The higher the value of resistance, the more "treble" in the sound since fewer higher audio frequencies are bypassed. The maximum bass "boost" is obtained when all, or almost all, the resistance is out of the circuit. In the case of Fig. 5-18A, this would be when the slide arm of the potentiometer is at one of its end positions. In the case of Fig. 5-18B, this same situation would prevail when the tone capacitor is connected directly to ground—when the switch is at the wire connection, rather than at R1, R2, or R3.

An alternate arrangement to the circuit shown in Fig. 5-18B above would be to have fixed capacitors of various values in place of R1, R2, and R3. A variable resistor would then be used instead of capacitor C1.

Sometimes, just as in the case of vacuum-tube ac-dc sets, you will find a capacitor hanging across the primary of the output transformer. This is a tone control of sorts but tone-compensation capacitor might be more nearly correct. Its function is to cut down the highs, thus diminishing the somewhat tinny sound produced by the speaker.

Figure 5-19 shows a gain-control circuit. The secondary of the interstage transformer, T1, is shunted by a potentiometer. This picks off the amount of audio signal that will be used to modulate the dc bias supplied by R1 and R2. Note C1, connected to the collector. This is a form of tone "control" but the word control here must be used tongue-in-cheek. C1 bypasses high tones to ground or chassis, giving apparent emphasis to lower-frequency tones.

THE EMITTER FOLLOWER (GROUNDED COLLECTOR)

In most audio amplifiers, the signal output is taken from the collector circuit. However, it is sometimes advantageous to take the output from a different element, such as the emitter. Such a circuit is known as an emitter follower or grounded collector. The emitter follower has a high input resistance, a low output resistance, and no phase reversal. Because of these characteristics, the emitter follower is useful in coupling to low impedance loads. The load can be another transistor, producing a rather nice circuit for us known as the direct-coupled amplifier. A circuit of this sort is shown in Fig. 5-20. Let's start with the collectors. The collector of the output transistor, Q2, is transformer coupled to the speaker. The primary resistance of the output transformer is fairly low, so the collector of Q2 gets almost the full negative voltage. The collector of Q1 looks as though it has a load resistor, R3, but this resistor has such a very small value that practically no signal develops across it. If it did, we could always bypass it with a capacitor shunted right in parallel with R3.

We can see that R1 and R2 act as voltage divider for Q1 but this isn't quite so clearly shown for Q2. If you will examine Q2 you will

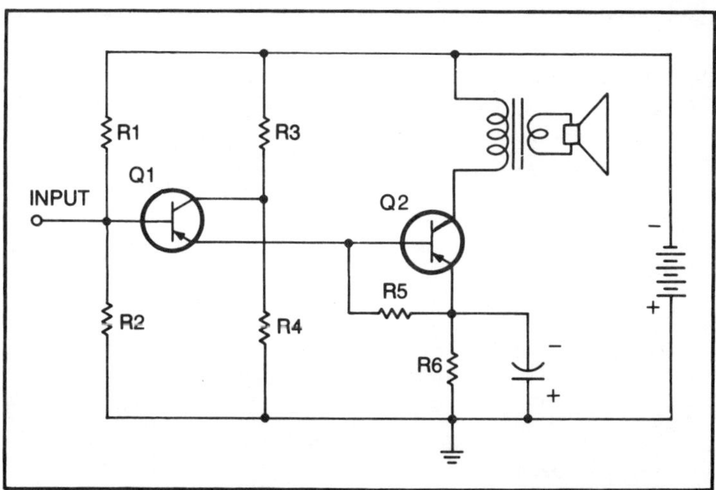

Fig. 5-20. The emitter-follower circuit in this arrangement is used as a driver. Q1 is direct-coupled to Q2. Q2 and its circuitry acts as the load on Q1.

see that it forms the emitter load for Q1. Thus, R1 and R2 really supply bias for both transistors.

How does current flow in a circuit of this sort? If you will start at the negative terminal of the battery, you will be able to move over to R3, through Q1 and then back to the positive terminal of the battery through the base of Q1 and R2. This base current is very small, so let's get back on to the main path. We can now go from the collector of Q1 over to the emitter of Q1. From here, following along with the emitter current, we can go through R5 and R6 or through the base-emitter circuit of Q2. But the emitter current of Q1 is our amplified signal current.

Now let's get back to the negative terminal of the battery once more. This time, suppose we move down through the primary of the output transformer, through Q2 and through R6 to ground (or the plus terminal of the battery). Is this the only current flowing through Q2? Not at all, since we just described the passage of Q1's emitter current through Q2. We now have two currents going through Q2. One is the steady direct current supplied by the battery. The other is the amplified signal current out of the emitter of Q1. This signal current will modulate the steady current of Q2. R6 is not only the stabilizing resistor for Q2, but R6, together with R5, form the emitter return path to ground for Q1.

What about the input to the direct-coupled stage? There is nothing unusual here. It could be resistance-capacitance coupled to the preceding circuit or transformer coupling can be used.

COMPLEMENTARY SYMMETRY IN DIRECT COUPLING

Transistors have one (some would claim more) great advantage over tubes. We got a glimpse of this advantage when we examined the complementary-symmetry push-pull amplifier. Complementary symmetry is just another way of saying that pnp and npn transistors work with opposite polarities and opposite directions of current flow. We've put this principle to work for us again, in the direct-coupled amplifier circuit shown in Fig. 5-21. The fundamental idea is the same, though, as in the earlier direct-coupled circuit we described. The difference is that this time Q2 forms the transistor load for Q1. The advantage here is that loading the collector produces more gain than the emitter follower. And it isn't necessary for the pnp transistor to act as the driver for the npn. The transistors can be transposed, but when we do, we must also remember to transpose the batteries.

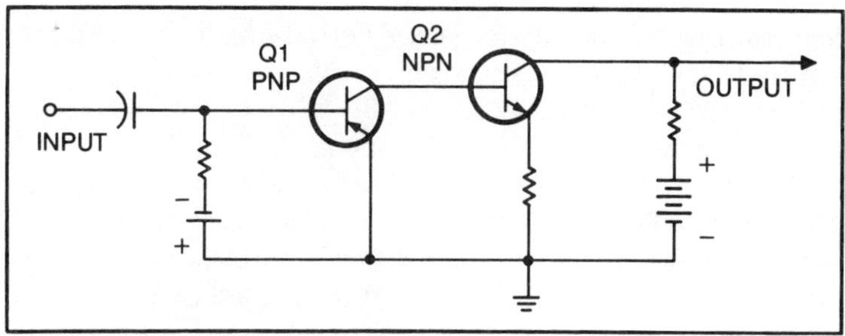

Fig. 5-21. Direct-coupled amplifier uses complementary-symmetry principle.

A FEW THOUGHTS

What is it that we are trying to do when we couple one transistor stage to the next? All we want is to vary the forward bias so that it goes up and down exactly in step with every little wiggle and movement of the signal. If you will look back at the single-ended stage shown in Fig. 5-10, and if you could change R1 into a potentiometer that you could operate rapidly enough, you would get audio output. This gives us a clue to how we can operate an audio stage, as shown in Fig. 5-22. Instead of having R1 and R2 as a voltage divider to supply forward bias, we have R2 and some kind of a transistor circuit substituted for R1. But is this so surprising? Earlier, in Fig. 5-19, didn't we put the secondary winding of an interstage transformer in series with R1 and R2? What we have in Fig. 5-22, then, is just another way of varying the forward bias.

What would a circuit of this sort look like? We've already had several, but may not have recognized them as such. As an example,

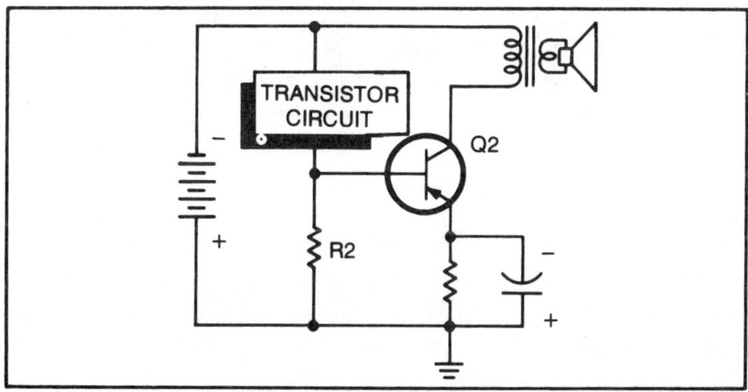

Fig. 5-22. A transistor circuit can be substituted for one of the forward-biasing voltage-divider resistors.

let's go back to Fig. 5-20 in which we show an emitter-follower circuit used as a driver. If you will examine the base circuit of Q2, you will see that we have R5 and R6 in series. One end of R6 connects to the plus terminal of the battery. And one end of R5 goes to the base of Q2. These two series resistors, then, correspond to R2 in Fig. 5-22. Now what about the block marked "transistor circuit?" Still working with the emitter follower of Fig. 5-20, we know that we must work our way up from the base to the other end (the minus end) of the battery. To do this we must go through the collector-emitter circuit of Q1 and R3. But the collector-emitter circuit of Q1 and R3 just represent a substitute for R1. And since this substitute supplies a varying audio signal, we will get audio modulation of the base circuit of Q2.

DIODE COMPENSATION OR STABILIZATION

All along we've been using a pair of series resistors, which we have marked R1 and R2, to supply forward bias for our transistors. But there is an assumption here that we haven't told you about. We have taken for granted that we will have a constant voltage drop across the resistors. Was this reasonable on our part?

To answer this question we need to ask another one. What could possibly cause the voltage across R1 and across R2 to change? Presumably a weak battery could be one cause of this trouble, but we can't take this too seriously. If the battery is weak, its regulation will be poor, the forward bias will be incorrect and we will get distortion. The end result of this is that what we hear out of the speaker will be a constant reminder to get a fresh battery. Now assuming that we have a fresh battery, what other difficulties could we have? What about our resistors, R1 and R2. As resistors, they have a positive temperature coefficient. But isn't this the opposite of what the transistor has? If you will examine Fig. 5-23A you will see that we have R2 in shunt with the base-emitter circuit of the transistor. This is like compounding a felony, since R2 and the base-emitter circuit have opposite temperature coefficients.

To get around this unhappy situation, we can substitute a diode for R2, as shown in Fig. 5-23B. The theory here is that the diode has a negative temperature coefficient. But what about the base-emitter part of our transistor? That's a diode, too.

Suppose temperature increases. The resistance between base and emitter decreases, causing an increased current flow between these two elements. At the same time, though, the resistance of the diode (substituting for R2) also decreases. Because the diode is in

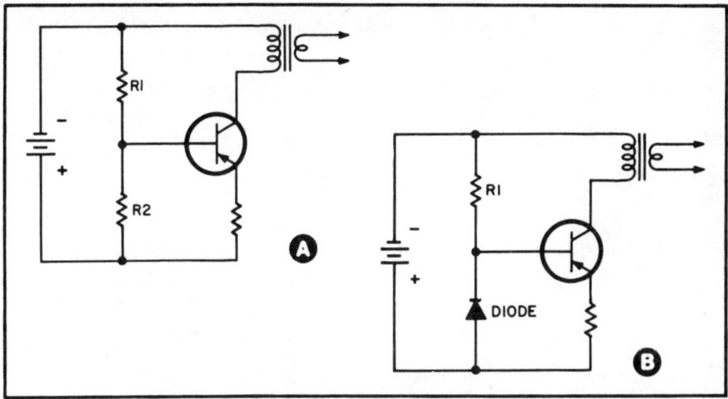

Fig. 5-23. The circuit shown in (A) uses R1 and R2 as a voltage divider for forward bias. In (B) we have substituted a diode for R2.

shunt with the base-emitter circuit, we have a bypass path for additional current.

Now what if we had a temperature increase and had R2 in place of the diode? With a rise in temperature, the resistance of R2 goes up and so this very action forces even more current to flow between base and emitter.

Diode compensation or stabilization is generally found with power output stages since it is in such stages that temperature has a chance to do its dirty work. In Fig. 5-24 we have a circuit using a

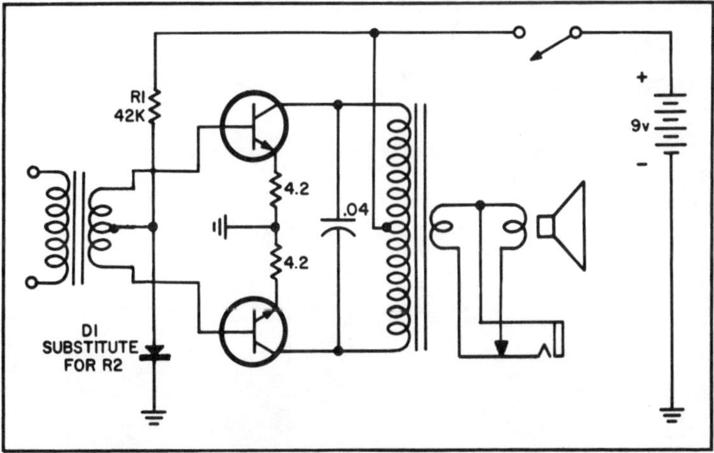

Fig. 5-24. Diode D1, substituted for voltage-divider resistor R2, has the advantage of acting as a temperature-compensating device. Note that the diode is forward-biased. Its forward-biased resistance is the correct value so that the bases of both transistors are properly biased.

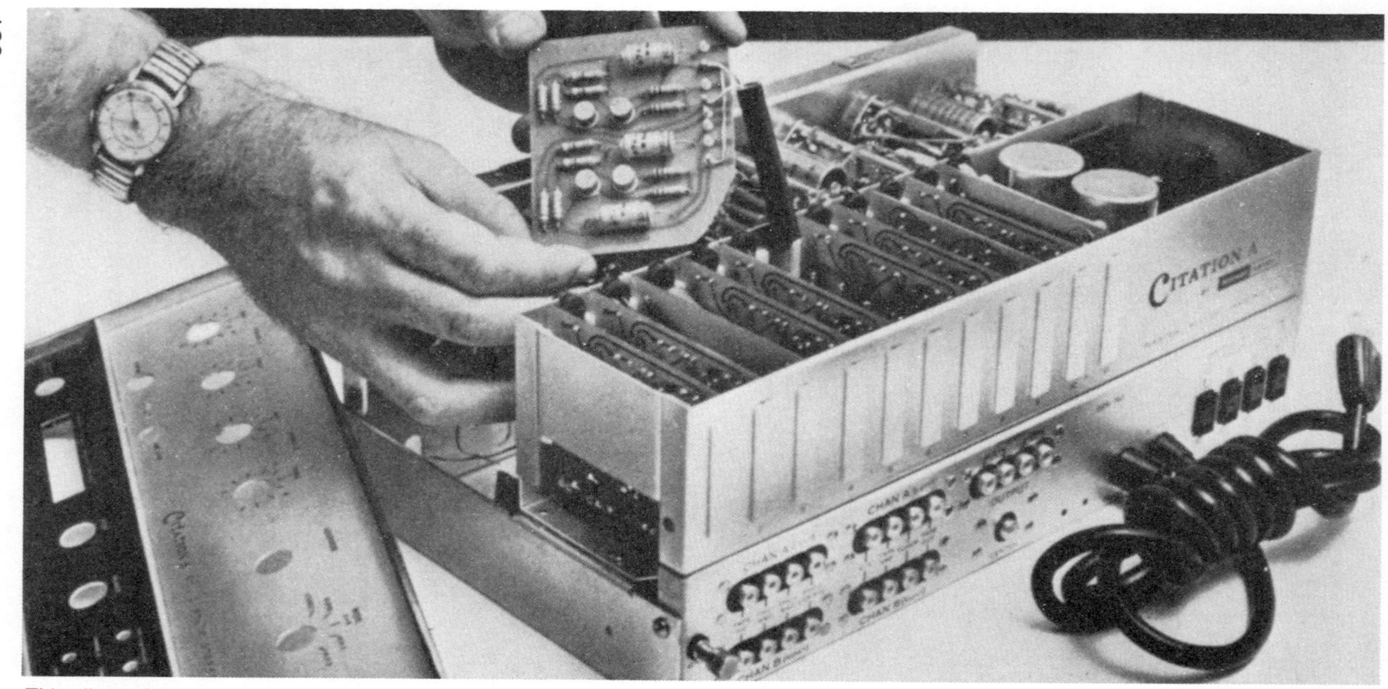

This all transistor stereo preamplifier features computer-like plug-in modules. A transformerless circuit gives a 1 hertz to 1 megahertz frequency response. (Harman-Kardon Inc.)

diode in this way. Note also the capacitor connected from collector to collector. This bypass is a bass-boost unit. The higher the capacitance, the higher the boost. A typical value would be .04 μF.

The circuit of Fig. 5-24 uses npn transistors. Pnps could also have been used, but in this case both the diode, D1, and the 9-volt battery would need to be transposed.

A FINAL WORD

Amplifiers, like people, come in all sizes and shapes. A popular indoor sport among engineers is to try to see just how many different variations of audio amplifiers they can dream up—and, considering the choice of audio transistor circuitry we now have, they haven't just been wool-gathering. We have push-pull drivers feeding push-pull output; amplifiers in which all transformers (interstage and output) are eliminated; new arrangements of bias supplies; various feedback arrangements; temperature compensation, etc. This is like a huge plate of spaghetti. Dig around with sufficient seriousness, and you're bound to find yourself a few meatballs. The same is true of any audio circuit. Keep in mind just what it is the transistor is supposed to do, rearrange the circuit so that the biasing becomes obvious, and you'll be able to brush aside the clutter of extra components with ease, so that the basic amplifier circuits just shine through at you.

This doesn't mean we're finished with amplifiers. Those we've examined cover audio frequencies. We now have to climb the frequency range up to rf. We start our ascent in the next chapter.

QUESTIONS

1. What is base current? Emitter current? Collector current?

2. What is a common-emitter circuit? Common-collector? Common-base?

3. What is meant by in phase? Out of phase?

4. What is phase inversion?

5. Describe biasing of a transistor amplifier. What is the effect of forward biasing? Reverse biasing?

6. Describe class-A, -AB, -B and -C amplifier operation.

7. What are the differences between voltage and power amplifiers?

8. How does a single-ended amplifier differ from a push-pull amplifier?

9. What are the advantages and disadvantages of R-C coupling?

10. What are the advantages and disadvantages of transformer coupling?

11. What is meant by temperature coefficient?

12. What is positive temperature coefficient of resistance? Negative temperature coefficient? Zero temperature coefficient?

13. What is a thermistor? How is it used?

14. How does temperature affect a transistor?

15. What is a heatsink? How is it used?

Chapter 6

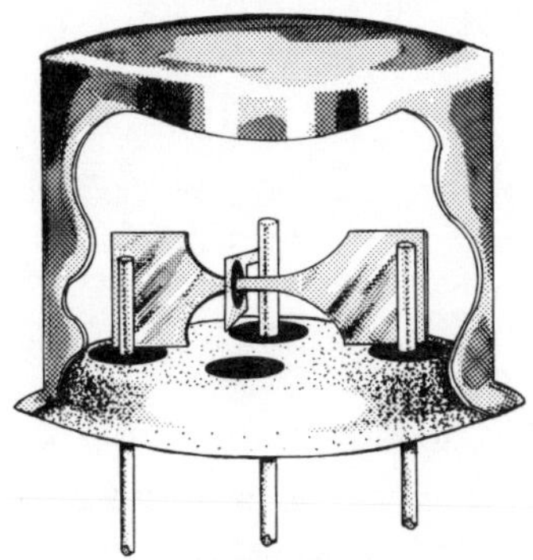

The Receiver

IF YOU LIKE THE "WHODUNIT" TYPE OF NOVEL, YOU KNOW THAT THE trick is to supply as many clues as possible and yet spring a surprise ending on the reader. In an electronics book of this sort we also supply as many clues as possible. The big surprise will come when you realize that basically, transistor circuitry is fairly simple and repetitive.

The audio amplifiers we studied in Chapter 5 live on a diet of dc and ac. The dc is supplied by one or more batteries. The ac is the signal, and the closer we get to the speaker, the hungrier the transistor gets for hefty input. To build the signal to a value large enough to satisfy the audio driver we generally need several i-f stages. Note that we did not include the detector, although this stage follows the i-fs as a matter of course. The reason for this is that in almost all transistor radio receivers the detector is a diode, and as such, supplies no gain.

TRANSISTOR DETECTORS

A transistor can be used as a detector and it does this job very nicely and easily, as we can see in Fig. 6-1. The circuit has several points of considerable interest, so let's take a good look at it. We can start with transformers T1 and T2. T1 is an i-f transformer and couples the transistor detector to the preceding stage. T2 is the audio-output transformer and is our connecting link to the following audio stage. T1, as shown by the short lines separating the primary

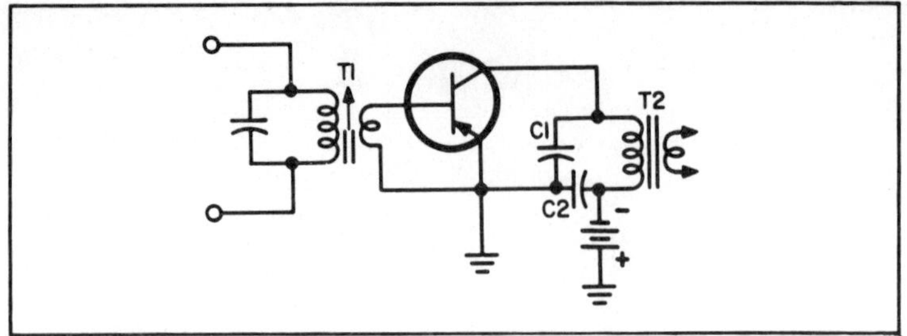

Fig. 6-1. Transistor detector works class B. In absence of a signal, the transistor is cut off. A signal of proper polarity supplies forward bias. The transistor rectifies, and supplies gain.

and secondary, has a ferrite core or slug. The arrowhead on top does not indicate the proximity of Indians, but rather that the core is tunable. Since we have a single arrowhead we know that this core will tune either the input or output side of T1, depending on its position. Sometimes you will see a double dashed line (each with its own arrowhead). This means that both primary and secondary are tunable. T2, the audio transformer, has a laminated iron core, and, of course, this is fixed.

We've been talking about i-f at T1 and audio at T2 and so, as you have probably gathered, something takes place in between these two transformers to effect this conversion. The change from i-f to audio is done for us by the transistor, working as a detector or signal rectifier.

THE CASE OF THE MISSING BIAS

What is so unusual about this transistor, though? We've seen transistors before. If you'll look carefully, you'll see that while we have the collector-emitter circuit properly biased by the battery, we have made no provision for biasing the base-emitter or input circuit. It is true that the emitter is tied to the plus terminal of the battery, but as far as the base is concerned, it is way up in the air.

The low resistance of the secondary winding of i-f transformer T1 might easily lead us to believe that the base is practically shorted to ground (or to the emitter). It is if we consider it only from the viewpoint of dc. But the impedance of the secondary is adequate for the signal.

EFFECT OF THE SIGNAL

Since we have no forward bias on the transistor, very little collector current flows. Suppose, though, that transformer T1 receives a signal voltage from the preceding stage, and let us also suppose that the polarity of this signal is such that the top end of the secondary is plus and the bottom end is minus.

What has the signal done? Because the secondary of the i-f transformer is connected between the base and the emitter, we now have a signal voltage between those two elements. The signal has made the base positive and the emitter negative. But this is reverse bias! Its effect is to drive the transistor collector current down toward the cutoff point. As a result we get no signal output and transformer T2 just sits idly by, waiting for something to happen. This won't take too long since the polarity of our input signal changes. The top end of the secondary of T1 now becomes negative and its bottom end positive. This is just what the base-emitter circuit has been waiting for, since this is forward bias. The collector current now climbs, its value depending on the strength and duration of the incoming signal.

DETECTION AND AMPLIFICATION

What has our transistor done? Two things! It has rectified the signal, since only half of the input waveform has any effect on the output. It has also amplified that half that did get through.

But what is the i-f? Isn't it just a sort of carrier with the audio superimposed on it? All we want for the audio amplifier is audio signal, though, the carrier being valueless at the output of the transistor detector. For this reason we put an i-f bypass (C1) between collector and emitter. Like all other capacitors, its reactance varies inversely with frequency. Since the i-f will probably be at least 455 kHz while the audio might have a top range of 10 kHz (if that high), you can readily appreciate that C1 will do its best to bypass the i-f but will offer considerably more opposition to the audio signal. The audio signal passes through the primary of T2, and promptly gets transferred, by electromagnetic induction, to the next stage.

DECOUPLING

This sounds so neat and pat that it may be hard to believe—and you would be right to be skeptical. Some audio does sneak through C1, thus getting back to the emitter, and never has a part in ultimately moving the voice coil of the speaker. And some i-f and audio does get through the primary of T2, down to the battery. To keep these currents from circulating through the battery, we protect ourselves with another bypass, C2. Since the battery we show in Fig. 6-1 is used by all stages, we can't have the battery working as a coupling device. When C2 is fresh, it just goes along for the ride. But batteries get older and as they do they are very much inclined to act as coupling devices in addition to their primary function of supplying a voltage.

Transistor detectors are used where the gain of the detector is essential. This is not the case in the average 6 (or more) -transistor radio, but it is sometimes used in inexpensive two-transistor receivers. We have the circuit in Fig. 6-2.

With two-transistor receivers we need every bit of gain, and so these are often made with a collapsible rod antenna. The tuned circuit is L1—C1. C1 is a single-section variable capacitor having a maximum capacitance value of 365 pf. R1, the volume control, is a 5-megohm potentiometer shunted across the primary of audio transformer, T1. Q1 works in the same manner as described for the detector circuit in Fig. 6-1. It would be helpful to have an rf bypass capacitor connected to the collector of Q1, but since sets of this sort sell for a very low price, this component is often omitted. Q1 gets its forward bias from the signal, and its collector-emitter bias from a 3-volt battery.

After detection and amplification by Q1 the signal is transferred by an audio stepdown transformer to the base-emitter input circuit of audio-amplifier transistor Q2. This transformer could have a primary impedance of one-hundred thousand ohms or more and a secondary impedance of about a thousand ohms.

Forward bias for Q2 is supplied by a single resistor, a 100,000-ohm unit, R2. This may seem a little strange, since we've become accustomed to a series-resistor voltage divider arrangement shunted across the battery for forward bias. We can examine

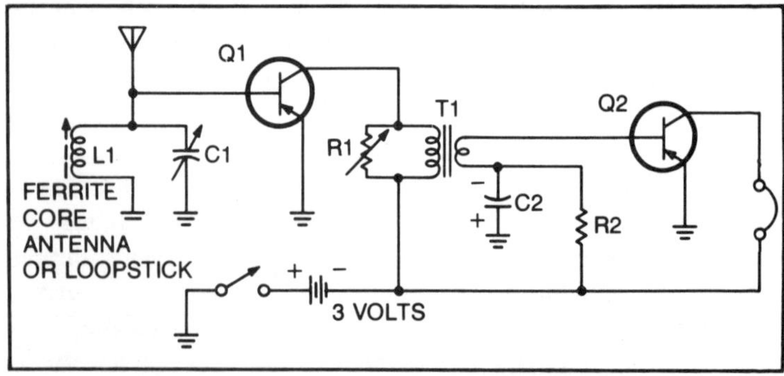

Fig. 6-2. Two-transistor receiver, Q1 is the amplifying detector; Q2 the audio amplifier.

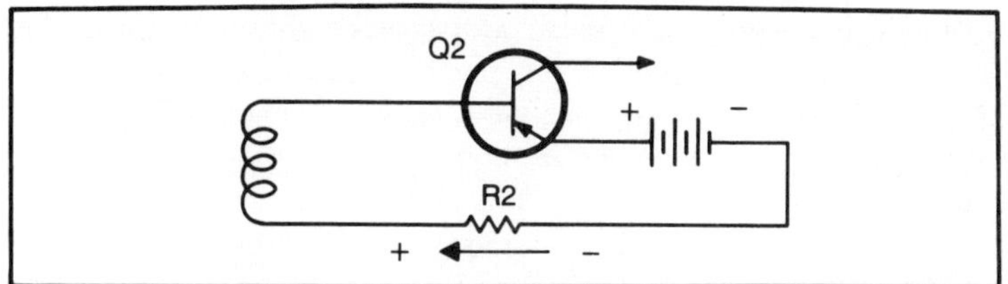

Fig. 6-3. Forward-biasing system for Q2 depends on a single series resistor.

the biasing arrangement by taking a closer look at the components involved, in Fig. 6-3. Instead of having the bias resistors in parallel with the battery, we have just a single resistor, and it is series connected. Again, the idea behind an arrangement of this kind is economy. Since we are working with a pnp unit, we want the base negative with respect to the emitter, but certainly not as negative as the collector must be. The arrow shows the direction of base-emitter current. This current passes through R2, but the voltage developed across R2 is in opposition to the battery voltage. The base-emitter voltage is equal to the battery voltage, minus the drop across R2. The signal voltage, appearing across the secondary of T1, modulates the current moving in this circuit, and as a result we get a similar (but amplified) variation in the output.

Getting back to Fig. 6-2, we have a capacitor (C2) rated at 3 working volts dc and 10 μf functioning as a bypass at the bottom end of the secondary of T1. The purpose here is to keep any audio frequency currents out of R2. We depend on R2 to furnish bias and we would like to keep the voltage across R2 as steady as we can.

AUTOMATIC GAIN CONTROL

Since transistors have been designed to supply gain, we can't get too exasperated when they amplify weak signals and strong signals with equal enthusiasm. It would be much better if the transistor showed a little judgment here, but that would be expecting too much . . . or would it? What do we want the transistor to do? We would like it to use its full gain for very weak signals, less gain for stronger signals; and for very strong signals not to flex its muscles at all. We force the transistor to do this with the help of a very simple circuit with not such a simple name. We call it automatic gain control, which we promptly abbreviate as agc. Sometimes the same circuit is called automatic volume control, or avc.

To understand agc, consider how the transistor works. As we increase forward bias, collector current increases. A strong signal will give us a much greater rise in forward bias than a weak signal. If we could increase the forward bias for a weak signal and reduce it for a strong signal, we might be able to rid ourselves of the annoyance of sound blasting from the speaker as we go from station to station. Agc isn't a cure-all, but it does help alleviate the problem somewhat.

Our first step toward getting agc is shown in the stripped-down circuit of Fig. 6-4. T1 is an i-f transformer and is responsible for delivering the i-f signal to the diode detector. As you can see, our diode detector is just a half-wave rectifier. R1, our volume control, is our diode load. When the signal across the secondary of T1 has the correct polarity (diode cathode negative) current will flow through R1 in the direction of the arrow. This makes the top end of R1 negative with respect to ground. It is true that the voltage across R1 is a varying audio signal, but it flows in one direction only, hence is a changing dc. We can take this dc voltage, and send it through a filter consisting of C1, C2, and R2 and by the time we get to the arrow marked agc, we have a nice, steady dc voltage.

But can we really boast about the steadiness of this dc voltage? If we get a strong signal, we will get a greater voltage drop across R1. This means more voltage for our filter. A weak signal means less voltage. We now have available a negative voltage (with respect to ground) whose strength is determined by the signal.

Let's see what we can do with this negative voltage. In Fig. 6-5 we have a transistor i-f stage and a diode detector. The top end of the diode load is positive and the bottom end negative (for part of the time) when a signal is received. We have inserted this resistor right in series with our two voltage divider resistors that normally supply

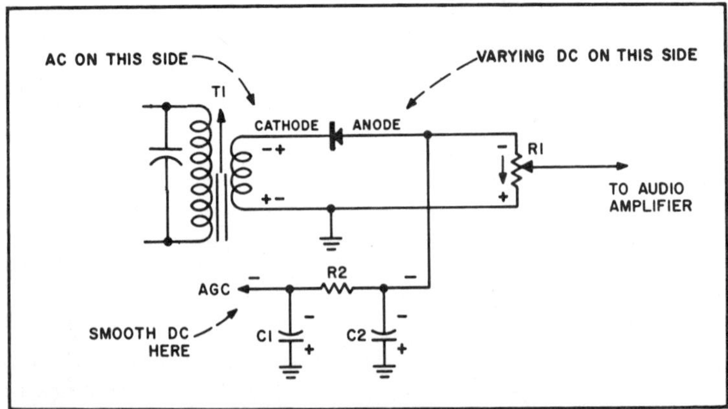

Fig. 6-4. Diode detector and agc circuit. R1 is the diode-load volume control. The voltage developed across R1 is also used for agc.

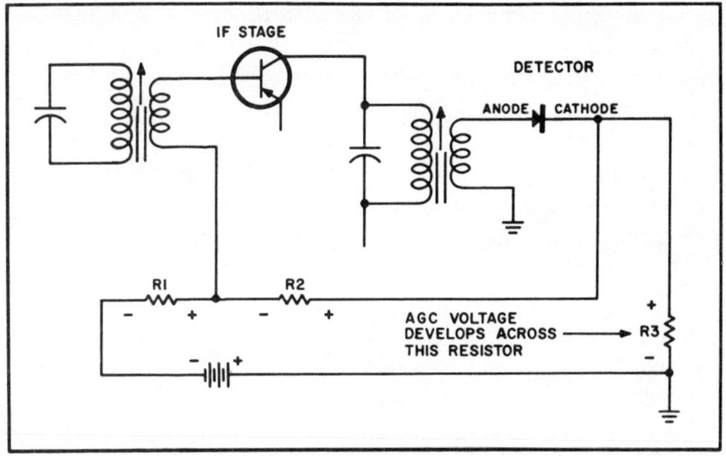

Fig. 6-5. Steps in the development of an agc system. The diode load is in series with R1 and R2. In the absence of a signal, forward bias for the i-f transistor is supplied by R1 and R2 in series with the diode load resistor. Resistors R1, R2 and R3 are shunted across the battery. Signal current, though, flows only through R3. The voltage developed across R3 is in opposition to the voltage produced across R3 by the battery current.

forward bias for the transistor. If you will examine the diode load, you will see that the voltage developed across it is in opposition to the voltage across R1 and R2. But how great is this opposition and on what does it depend? A strong signal will develop a stronger voltage across the diode load. But this stronger voltage will oppose the forward biasing arrangement of R1 and R2. As a consequence, the gain of the transistor in the last i-f stage (in Fig. 6-5) will not be as great as it would be without this opposition. A weak signal, of course, will produce much less voltage than a strong one, and so the transistor will be allowed to work with greater gain.

Now you might think this is a wonderful idea and just jim dandy, but look hard enough and you will find a price tag cached away somewhere. And where is the serpent in the agc Garden of Eden? The gain is reduced for all signals. We can tolerate this for strong signals, but for some weak signals it is tantamount to signal suicide.

As a matter of common practice, agc is often tapped from the top of the volume control (which acts as the diode load) and is fed back as a control voltage to the transistor in the first i-f stage. We have this arrangement in Fig. 6-6. The series arrangement, R1, R2, and the volume control, R3, act as the voltage divider for the first i-f transistor. Capacitors C1 and C2 are agc filters, bypassing any signal variations to ground or to the emitter of the first i-f. C4 looks as though it might be part of the agc network, but its function is that of an rf bypass, to keep the i-f out of the volume control, hence out of the audio system which follows. Similarly, C3 in the first i-f stage if also an rf bypass.

DECOUPLING

All of the stages in a transistor radio receiver, no matter what we may ask them to do, have a common denominator. The one component through which all the transistors are related is the battery. Transistors, though, have signal currents as their stock in trade. And these signal currents do not regard a battery with the same uncritical eye as a transistor-radio fan. To a signal current a battery is an impedance. When the battery is fresh, it has such a low impedance that the signals just go scooting right through it as fast as a junior scout on a greased pole. But as a battery gets used, or if it just sits around waiting to be used, its impedance rises just as surely and inevitably as your yearly taxes. When the impedance gets high enough the signal has a field day, since it can develop a signal voltage across the battery.

The battery, though, is common to all stages, and when in a somewhat discharged condition, acts as a coupling element. The sum and substance of all this is that there is nothing the signal

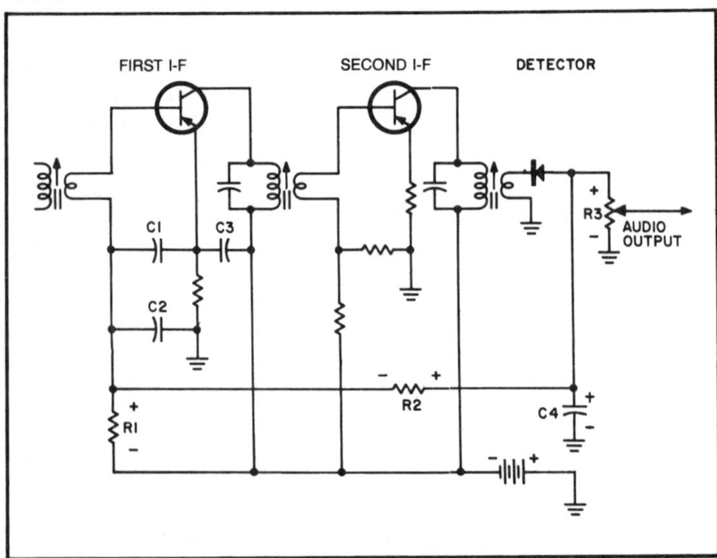

Fig. 6-6. Capacitors C1 and C2 and volume control R3 form the essential elements of the agc system.

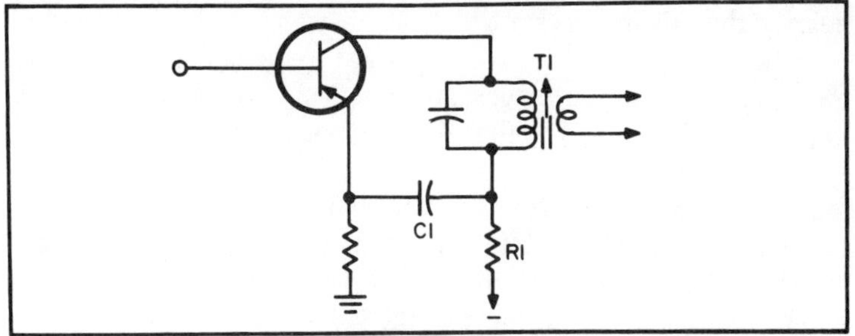

Fig. 6-7. R1 and C1 are the decoupling components.

currents would rather do than go wandering through the battery and then through the various receiver stages. Now we cannot have this since it is a form of signal anarchy. Once we lose control of the signal anything can happen, with the result usually evidenced as a series of weird noises out of the speaker.

We can avoid this unhappy state of affairs by shunting the battery with an electrolytic capacitor or by decoupling the collectors of the transistor. All that is needed are some series resistors and capacitors. A typical circuit is shown in Fig. 6-7. C1 is the bypass and R1 is the series resistor. R1 can have a value of several hundred ohms up to about 1K. A value of .05 μF for C1 is common. In this circuit, T1 is an i-f transformer.

AUXILIARY AGC

Nothing delights the heart of a design engineer more than his ability to make one component do two jobs. An emitter resistor, for example, not only supplies the correct voltage for the emitter, but (when its shunting capacitor is omitted) some negative feedback as well. In Fig. 6-7, decoupling resistor R1 can be made to do an interesting variety of chores. It works, as we have mentioned, as a decoupling resistor. It drops the battery voltage so that the correct amount is placed on the collector. And it contributes its little mite to the circuit shown in Fig. 6-8, and known as an auxiliary agc.

PROBLEMS, PROBLEMS

Now you might have thought that we were finished with agc. We are, except for a problem we have when operating a transistor radio in strong signal areas. Transistor receivers can produce an inspiring amount of gain and if we begin with a strong signal, it is entirely possible that the agc system couldn't cope with this combination. To help matters, we use the circuit shown in Fig. 6-8. The heart of this auxiliary system is the diode which we have connected from the collector circuit of one i-f stage to the collector circuit of the next. The two i-f stages are identical, except for one very important difference. The first i-f is not agc controlled; the second i-f is. Suppose, now, that no signal is being received. Without signal, the collector currents of the two stages will be the same (assuming identical transistors and components). If R1 and R2 are resistors of the same value, the voltage drops across these resistors will be the same. This means that the voltage at point A in Fig. 6-8 will be the same as the voltage at point B. Forgetting the resistances of the i-f transformer windings for the moment, this means that the cathode of the diode and its anode are both at the same potential. Under this condition, the diode might just as well not be in the circuit. However, we do have a little bit of a voltage drop across the primary of T1, and so we have a potential impressed across the diode. The polarity of this potential, though, is such that the diode is reverse biased. We need this touch of reverse biasing, since the diode, hanging on to the collector of our first i-f, is in a fine position to sabotage the signal. Reverse biasing, hence increased resistance, prevents this.

SIGNALS—WEAK AND STRONG

Now suppose we receive a weak signal or one that is just moderate. The effect of this is to reduce the forward bias of the second i-f. But if we do this, the collector current of the second i-f is

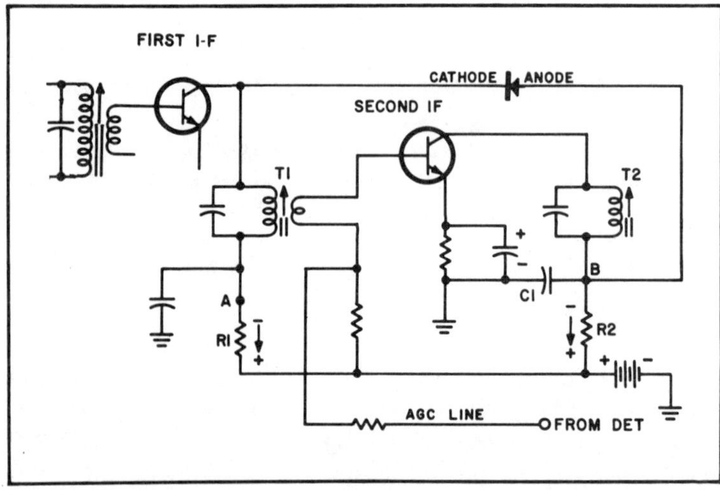

Fig. 6-8. This auxiliary agc system uses a diode shunted across an i-f stage.

also reduced. The collector current, though, flows through R2. This reduced current means less voltage across R2, with the result that point B becomes more positive. For weak or moderate signals, the increase of positive voltage at point B isn't too significant. but what if we get a signal that is strong? The drop across R2 will become much less. It will be just as though we had moved point B right down to the plus terminal of the battery. But this is connected to the anode terminal of the diode. The diode becomes forward biased.

From here on events take place rapidly. A forward biased diode acts very much like a resistor of very low value. When the signal arrives at the top of T1 it finds to its surprise that a shortcut awaits—an easy path to ground through the diode over to point B and from point B through the bypass capacitor (C1) to ground. Of course not all of the signal is killed, enough of it is getting through T1 and the second i-f to produce a respectable amount of volume out of the speaker. But no blasting and no mad dash to get at the volume control.

IMPEDANCE MATCHING

Every i-f transformer has two requirements, but unfortunately these demands can be in conflict—and usually are. One essential of an i-f winding is that it should have enough turns so that it can be tuned to the intermediate frequency. To be able to do this, it must have a certain number of turns. But as the number of turns is increased, so is the overall impedance of the winding. But we want the impedance to match the base-emitter input or the collector-emitter output.

Although this idea is fairly evident when we work with resistors, it isn't quite so obvious with inductors. To make sure we understand it, let's move on to Fig. 6-9. In Fig. 6-9A we have a pair of 500-ohm resistors in series, for a total of 1,000 ohms. The resistance from point A to ground is 1,000 ohms, but if we move the ground point up, as in Fig. 6-9B, the resistance from point A to ground is only 500 ohms. This is true even though we still have 1,000 ohms connected into our circuit. Now move over to Fig. 6-9C where we have a coil having an impedance of 1,000 ohms from point A to ground. If we put the tap at the *electrical* center of the coil (this is not the same as the physical center), then the impedance from point A to ground (Fig. 6-9D) is only 500 ohms. We do not need to connect a direct ground, for in Fig. 6-9E we see that we have a capacitor wired into our electrical centertap. This puts the center at ac ground potential and so the impedance from A to the point of

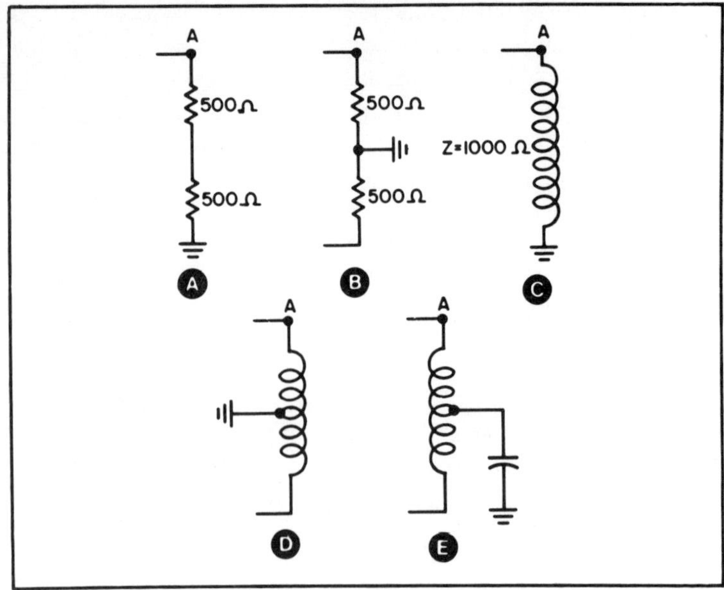

Fig. 6-9. Out of the total impedance supplied by a coil, we can select as little as we need for our transistor circuit.

connection for the capacitor is only 500 ohms. However, the entire coil might still be connected into our circuit.

All of this has a very practical bearing on our transistor circuits. We can take advantage of the techniques shown in Fig. 6-9 so that we get a good match between the transistor and the i-f transformers. If you will look at Fig. 6-10 you will see that we have the final i-f stage of a transistor receiver. The total impedance of the primary of the i-f transformer is measured between points A and B. Point B, though, is connected to C1, a 50 μF capacitor. One side of this capacitor is grounded and so, as far as any signal voltage is concerned, point B is at ac or signal ground potential. The collector is connected to a tap on the primary. The total impedance presented by the primary winding to the transistor is the impedance of the coil between the collector tapping point and point B. If we were to move the tap further up on the coil, we would get more and more impedance into the collector circuit. The maximum would be if we connected the tap to point A.

THE IMPEDANCE DIVIDER

Impedance division, such as that obtained through the use of a tapped coil, is no different from using a resistor as a voltage divider.

Actually, a tapped coil used as an impedance divider could also be called an ac voltage divider, a somewhat better name since this is a constant reminder that the impedance or voltage division is effective for ac only.

Looking at Fig. 6-10 again, you might get the impression that since the collector is tapped so far down, there is a large loss of signal voltage. The tap, though, has the effect of converting the primary into an autotransformer, supplying both a voltage stepup and a very nice example of having our cake and being able to munch on it as well.

Tapped coils can be used in the transistor input, the transistor output, or both, as shown in Fig. 6-11. It is applicable to both types of transistors—pnp and npn. And although we have indicated impedance matching by having the collector tapped down on the i-f transformer, we could connect the collector to the top of the coil and tap up with the emitter. This is shown in Fig. 6-12. The tap on the coil is at signal ground. The impedance presented to the collector is between the tap and the top end of the coil. Not only is the tap important for impedance matching purposes, but its position also helps control the amount of negative feedback, when this is used.

NEGATIVE FEEDBACK OR NEUTRALIZATION

If the transistor is behaving itself, the signal in the collector circuit is much more substantial than the signal in the input of the same transistor. Do you remember the story of Lot's wife? As told in the Bible, she looked back and was turned into a pillar of salt. Now this won't happen to our output signal but we want it to look forward, not backward. We don't even want a small fraction of it to go backward, because if it does, it may cause miscellaneous trou-

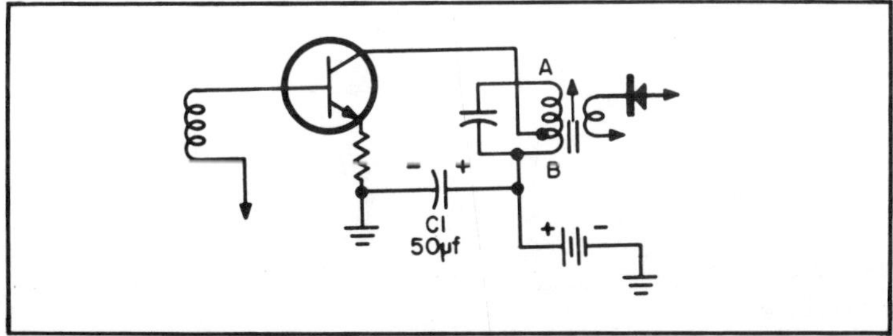

Fig. 6-10. The collector of the transistor is tapped down on the primary of the i-f transformer to obtain an impedance match. Capacitor C1 acts as an ac short to ground. Maximum impedance would be at point A on the coil winding. The impedance gets less as we move toward point B.

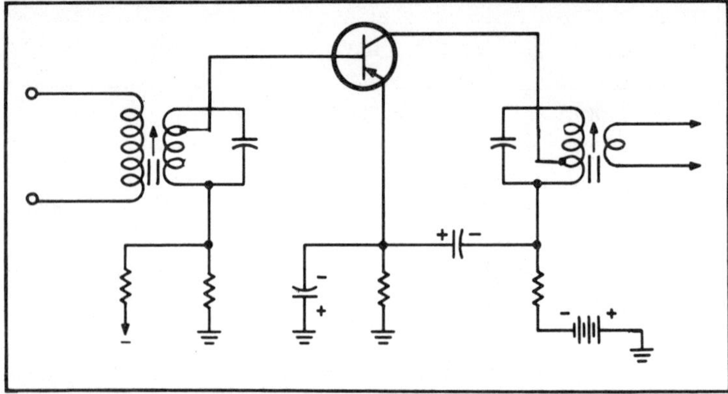

Fig. 6-11. Both the base and the collector are tapped down on their respective i-f transformers for impedance matching.

bles such as oscillation and narrowing of the bandpass. We're particularly worried about the i-f stage since it so happens that its design is just ideal for signal feedback. Our transistors are triodes and our input and output circuits are tuned to the same frequency. Triodes love to oscillate and do so with carefree ease, especially with i-f transformers. Now the very strange thing about all this is that for oscillation to take place, the fed back signal (the output signal) should have the same phase as the input. But in the common emitter circuit, the output is out of phase with the input. However, between the stray capacitance in the circuit, and the inductance of the i-f transformer, we can get enough of a phase shift so that the output signal drifting back to the input is sufficiently in phase with it so that it creates mischief.

The answer to this problem (see Fig. 6-13) is to feed back,

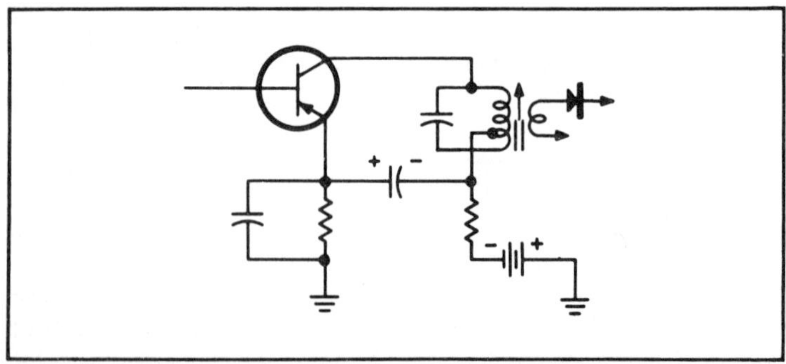

Fig. 6-12. For impedance matching, we can have the collector tap down (as in Fig. 6-11) or the emitter tap up on the i-f winding. The result is the same.

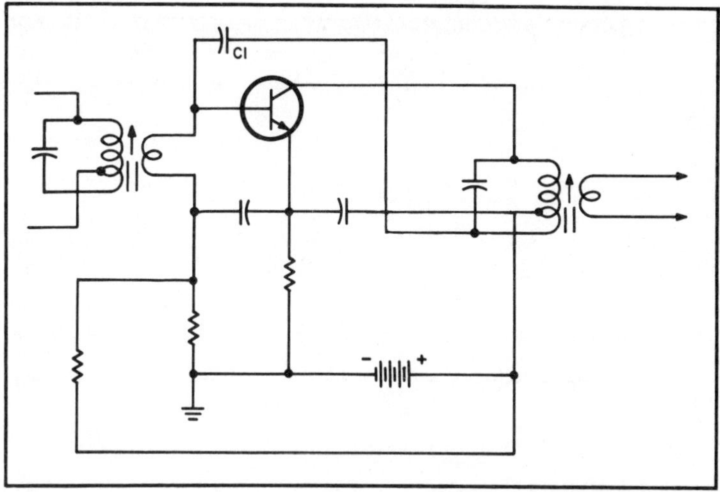

Fig. 6-13. C1 is the neutralizing capacitor. The amount of negative feedback depends on the value of C1 and the position of the tap on the primary winding of the i-f transformer.

deliberately, enough of the output signal to the input, to cancel the signal leaking back accidentally. We connect a feedback capacitor, C1, between the output and the input. Note that C1 is really attached to the bottom end of the coil while the collector bias is attached to the tap on the i-f transformer. This provides an out-of-phase signal to cancel the collector-to-base feedback in the transistor. We can adjust the amount of neutralization feedback by moving the tap, but remember, when we do so, we are also changing the impedance of the transformer and this is also the collector load. Changing the value of C1 will also change the amount of feedback to the base circuit. Putting resistance in series with the capacitor will reduce the feedback. The best way to control this is with a variable capacitor that can be adjusted so that we get just the amount of feedback that is needed. This is the most expensive technique.

FEEDBACK VARIATIONS

There are other ways in which we can neutralize a triode transistor so that it does not oscillate. A few of these techniques are shown in Fig. 6-14. In Fig. 6-14A we have a capacitor connected as in Fig. 6-13. The capacitor can have a very small value, a few micromicrofarads being sufficient. We can also cut down on the amount of feedback by putting a resistor in series with the feedback capacitor as in Fig. 6-14B. It makes no difference whether the resistor comes before or after the capacitor (Fig. 6-14C). We can also make our feedback connection directly to the input of the following stage, as in Fig. 6-14D. When neutralization is continued stage to stage the circuit in 6-14E is used.

THE PROBLEM OF SELECTIVITY

Every receiver—whether transistor or tube—has at least one variable tuned circuit and a number of fixed tuned circuits. We want tuned circuits in a receiver because we need them to be able to separate one station from the next. Variable tuned circuits are those

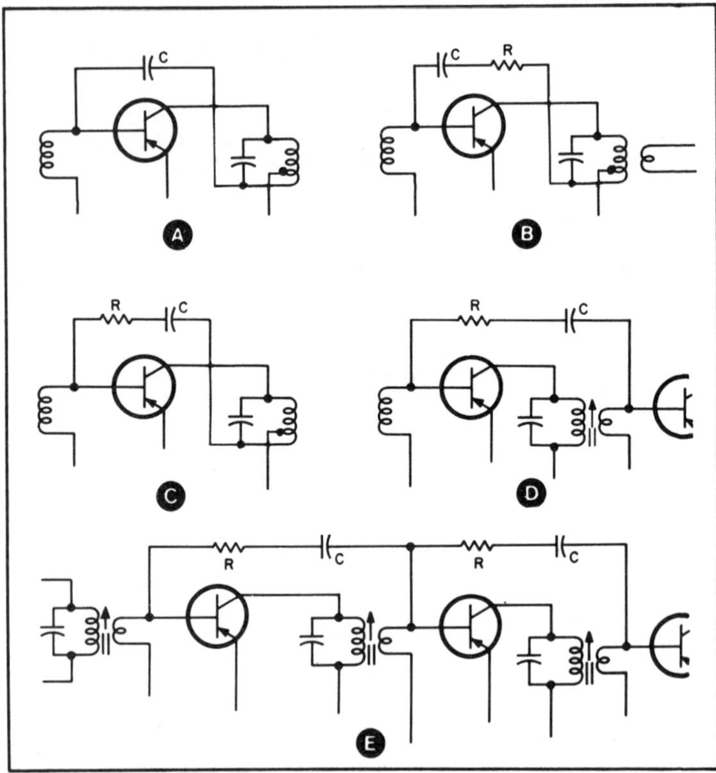

Fig. 6-14. Some negative feedback circuits for neutralization of i-f amplifiers. The simplest method (A) uses a fixed capacitor and a tapped transformer. A series resistor can be inserted as in B to help control the feedback. The positioning of the resistor (R) and the capacitor (C) in B and C does not affect the results of the feedback. Sometimes the negative feedback is arranged from the base of the following stage (as in D) to the base of the preceding stage. The amount of feedback is controlled by R and C. Feedback is usually needed in more than one stage and is connected as in E when tapped transformers are not used.

that are continuously adjustable by either a moving ferrite slug or a variable capacitor. A fixed tuned circuit, as its name implies, is tuned to a particular frequency and remains that way. Theoretically, variable tuned circuits are just dandy, but in practice having more than one tuned circuit presents a host of practical problems, particularly the very serious one of having all of the tuned circuits remain in step throughout the whole tuning range. This was the difficulty faced by early radio receivers which invariably were of the trf (tuned radio frequency) type. At first each circuit was tuned separately. This meant a separate tuning dial for each circuit and it was not uncommon to see radio sets with three dials, each one of which had to be adjusted whenever a different station was to be tuned in. As you can imagine, tuning from one station to the next was not the most popular indoor sport. Then some bright-eyed lad had the brilliant inspiration of putting all the tuning capacitors on a single shaft. This worked fairly well, but in general it limited the trf to about three tuned circuits.

The superheterodyne changed this, for each tube-type i-f stage could supply two tuned circuits. If a receiver had three i-f's, this meant a total of six tuned circuits, thus supplying at once twice as many tuned circuits as the old trfs.

FREQUENCY CONVERSION

Briefly, a superheterodyne works on the principle of frequency conversion. The receiver has a variable tuned circuit capable of selecting any frequency on the broadcast band. In the receiver, a circuit known as a local oscillator also produces a signal. We now have two signals, one supplied by the broadcast station and the other, locally, by the receiver. When these two signals are mixed, the result is a third signal which we call the intermediate frequency. The receiver oscillator (or local oscillator) is continuously adjustable, so that the difference between its frequency, and that of the radio station, is always constant. Thus, if you tune a radio set to a station at 1400 kHz, the local oscillator will be set by that same tuning action to 1855 kHz. When these two frequencies, 1400 kHz and 1855 kHz are mixed, one of the resulting frequencies will be the difference between the two, or 455 kHz. This is our intermediate frequency or i-f. If we tune our set to 800 kHz, our local oscillator will be at 1255 kHz. Once again, the difference between the two frequencies will be 455 kHz. Thus, the whole idea—the basic idea—of a superheterodyne receiver is to change any frequency on the broadcast band to a single fixed frequency. Because of this action, we can have many tuned i-f stages—actually fixed tuned because they remain tuned to but one frequency.

CONVERTERS

The job of changing the broadcast signal to an intermediate frequency (i-f) signal can be done by a pair of transistors or by a single unit. If two transistors are used, one is known as the mixer and the other is called the local oscillator. In practically all transistor receivers, however, just a single transistor is used to work both as a mixer and local oscillator. This transistor is called a frequency converter, or, more simply, a converter. If a separate transistor is used as the local oscillator, and a separate transistor as the mixer, we refer to the combined operation as a heterodyne action. If a single transistor combines both functions, we call the operation autodyne. Actually, the behavior is the same whether we use one transistor or two.

THE AUTODYNE CONVERTER

The converter transistor circuit receives the signal from the antenna at its input; delivers an i-f signal at its output. The trouble with this statement is that it is notable for the information it does not give us. The converter is a triple header. It:

1. Amplifies the incoming or rf signal.
2. Mixes or heterodynes the rf signal with the locally produced oscillator signal.
3. Amplifies the i-f signal.

This is quite a multiplicity of jobs for just one transistor, so let us separate these transistor functions and examine them separately.

THE LOCAL OSCILLATOR

An oscillator, as used in a radio receiver, is a radio or high-frequency ac generator. Typically, the oscillator works because we take a portion of the amplified signal developed in its output and feed it back to the input. Now this sounds very much like the neutralization techniques we described earlier, and so it is, with one basic difference. Neutralization uses feedback that is out of phase with the input; oscillation uses feedback that is in phase. These differences are shown in Fig. 6-15.

THE HARTLEY OSCILLATOR

In Fig. 6-16 we have a very popular oscillator; the Hartley. The

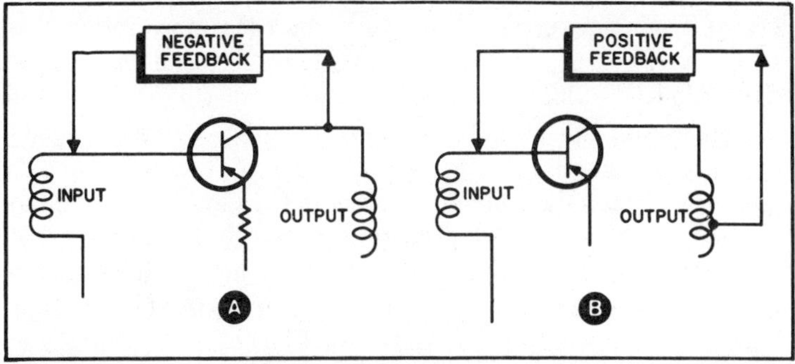

Fig. 6-15. We have two types of feedback: negative (also known as degenerative, inverse, or out-of-phase) for neutralization, and positive (also known as regenerative or in-phase). Positive feedback is used to produce oscillation; negative feedback to prevent it.

battery, resistors, and capacitors have been omitted for the moment. In this circuit current will flow from the collector to the emitter. The emitter is grounded and so is a tap on the coil. The current will flow through this common connection, up through the coil, back to the collector, thus completing its rounds. But the current, in passing through that part of the coil we call L1, built up a magnetic field around L1. This magnetic field induces a voltage across L2. If you will trace the connections to L2 you will see that it is in the base-emitter or input circuit.

Now let us suppose that the voltage across L2 has a polarity such that it supplies forward bias for the input. The result of this will be an increase in collector current. But if we get an increase in collector current, we will get a stronger growing magnetic field around L1, and this, in turn, will induce even more voltage across L2. This increases the forward bias, which, once again, produces more collector current.

Of course there is a limit. The output circuit cannot increase its current output indefinitely, and soon the collector reaches satura-

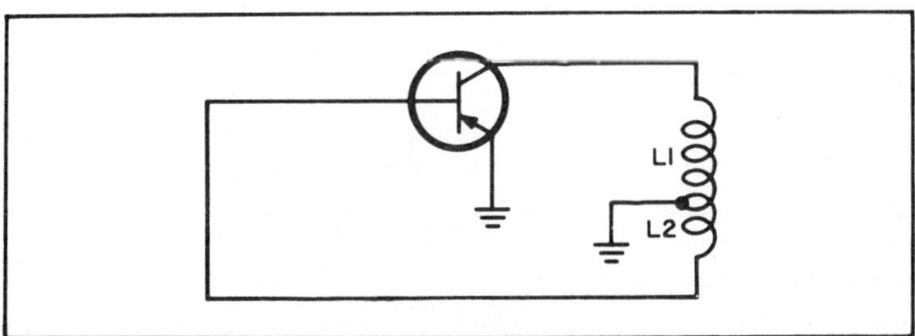

Fig. 6-16. Basic elements of a Hartley oscillator.

tion, or maximum current level. As it approaches saturation, the rate of change of increase in current becomes less and less. But to get an induced voltage we need a *changing* magnetic field—our induced voltage across L2 depends on it for its existence. Consequently, we get a smaller and smaller induced voltage across L2, until finally, when the collector current is at saturation, we get no induced voltage at all. At this point we have maximum current through L1 and a maximum magnetic field around it. But this magnetic field is a steady one and as such, is incapable of inducing a voltage. What has happened to our forward bias? Since it consists of the voltage across L2, it has disappeared. But with no forward bias, the collector current starts to decrease. This decreasing current once again produces a changing magnetic field around L1 and once again we get a voltage induced across L2. But the polarity of this voltage is exactly the opposite of what it was before. Now the voltage across L2 reverse biases the input. This drives the collector current down to cutoff. But at cutoff, there is no current through L1. There is no magnetic field around L1 and we have nothing with which to induce a voltage across L2.

But isn't this where we came in? When the reverse bias is removed from across L2, a small amount of collector current begins to flow. This produces a growing magnetic field across L1, inducing a forward-biasing voltage across L2—and so the entire action repeats.

OPERATING FREQUENCY

In this circuit, the collector current runs the gamut from cutoff to saturation. The frequency with which it does this is determined by the value of L1. If we put a capacitor in shunt with L1 or in series with it, the frequency will be determined by the value of both components—the coil and the capacitor, the combination often being referred to as a tank circuit. L2 is sometimes called the tickler or feedback coil.

THE ARMSTRONG OSCILLATOR

Although they may not seem so, coils L1 and L2 in Fig. 6-16 form a transformer. This isn't immediately apparent since most of us are accustomed to thinking of transformers as units having separate primaries and secondaries. Transformers with a tapped winding, as in the case of the Hartley oscillator, are called autotransformers. In Fig. 6-17 we have two examples of such transformers. Figure 6-17A is a step-up unit; that in Fig. 6-17B is a

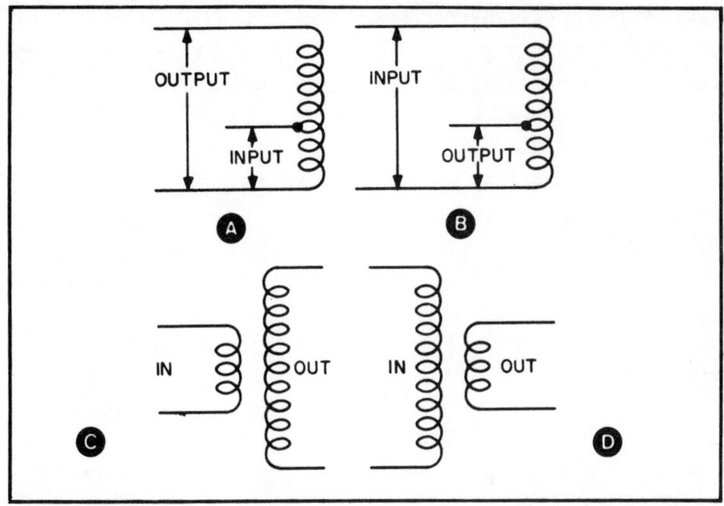

Fig. 6-17. Autotransformers and transformers with separate windings. In A we have a stepup autotransformer; its counterpart is shown in C. B is a stepdown autotransformer; its counterpart is shown in D.

step-down. Corresponding to these autotransformers we have those with separate primaries and secondaries, as in Figs. 6-17C and 6-17D.

The Armstrong oscillator uses a transformer having individual primary and secondary windings. See Fig. 6-18. The current works in exactly the same way as the Hartley. As far as any current variation is concerned, the two coils L1 and L2, are really con-

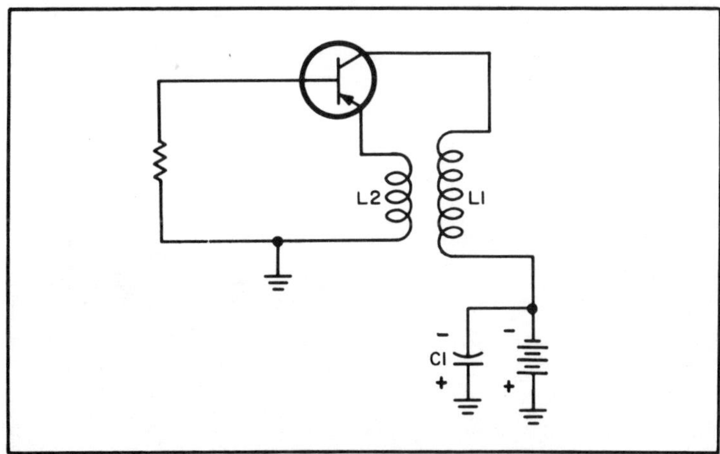

Fig. 6-18. Basic elements of an Armstrong oscillator.

nected. The bottom end of L2 is grounded. The bottom end of L1 is grounded through capacitor C1. This means that L1 and L2 are connected electrically, although not physically, as in Fig. 6-16.

THE COMPLETE OSCILLATOR

What we have shown you so far are the basic essentials of an oscillator. The local oscillator, though, must have a variable tuning circuit so that its operating frequency can be continuously adjusted when selecting a station. In Fig. 6-19 we have a variable tuning capacitor, C2, shunted across L2. This capacitor is mounted on the same shaft as the rf tuning capacitor (not shown in Fig. 6-19). Thus, when the rf tuning capacitor is turned to select the station desired, C2 is also turned at the same time. The local oscillator frequency is determined by the values of L2 and C2, and this frequency, as we mentioned earlier, is 455 kHz higher than the frequency of the broadcast signal.

Figure 6-19 has a few other items of interest. Note our old friends, R1 and R2, once more in the act. In this case, we choose a value of bias so that our transistor works in the low-current region. This is the part of the collector characteristic that is curved or non-linear. We must operate this way to get the local oscillator and rf (or broadcast) signals to beat or mix with each other.

R1 and R2 put a dc voltage on the base. But the base is connected to a tap on L2, one end of which is connected to ground. To keep L2 from shorting R1, we include a capacitor C1.

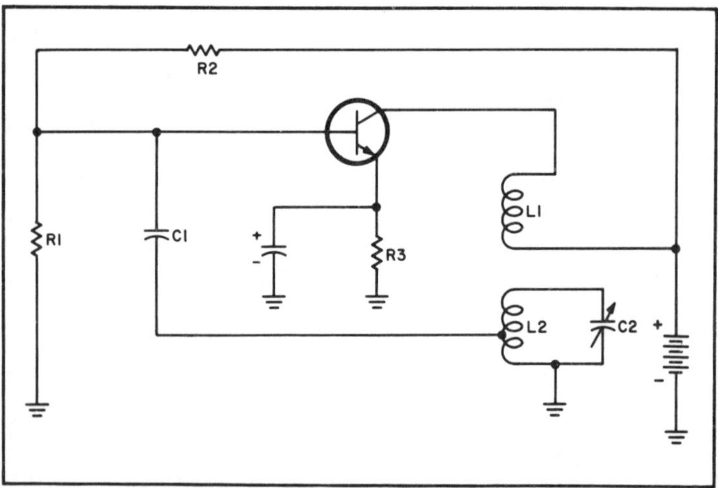

Fig. 6-19. Local oscillator circuit. L1 is the tickler or feedback coil.

THE MIXER

Now that we have our local oscillator, let's remove the transistor and see how it also works as a mixer. We're going to use the same transistor since it is going to operate both as a mixer and local oscillator. The circuit is shown in Fig. 6-20.

As you can see, the arrangement looks just like a straightforward amplifier. The rf signal is tuned by C1, shunted across the primary of T1. The rf tuning capacitor is mounted on the same shaft as C2 in Fig. 6-19. The signal is delivered to the base-emitter input circuit by T1. R1 and R2 are the same forward-biasing components we used in Fig. 6-19. R3 is also the same. In the collector circuit, though, we now have T2, our first i-f transformer. T2 delivers the i-f signal to the base of the following stage.

When the signal is received by T1 it is rf. T1 is a ferrite-rod antenna and picks up the signal. The signal that is delivered by T2 is i-f, and so, somewhere between T1 and T2, we must have some sort of frequency conversion, to change the signals of all stations on the broadcast band, to the single frequency required by the i-f stages.

Our final step to the complete mixer is to combine the circuits of Fig. 6-19 and Fig. 6-20, the local oscillator and the mixer, into just one circuit. We have this in Fig. 6-21, and if you will examine it carefully, you will see just how this was done.

Most of our components remain the same. C1 is ganged to C2, as shown by the dashed line connecting them. The oscillator transformer consists of L1 and L2. The primary of the i-f transformer, T2, is in series with L1 of the oscillator section. We've added some

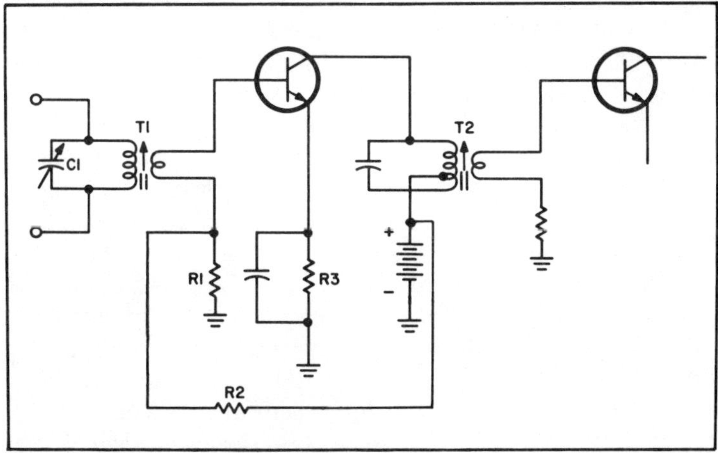

Fig. 6-20. Mixer circuit. The local oscillator has not been connected to it.

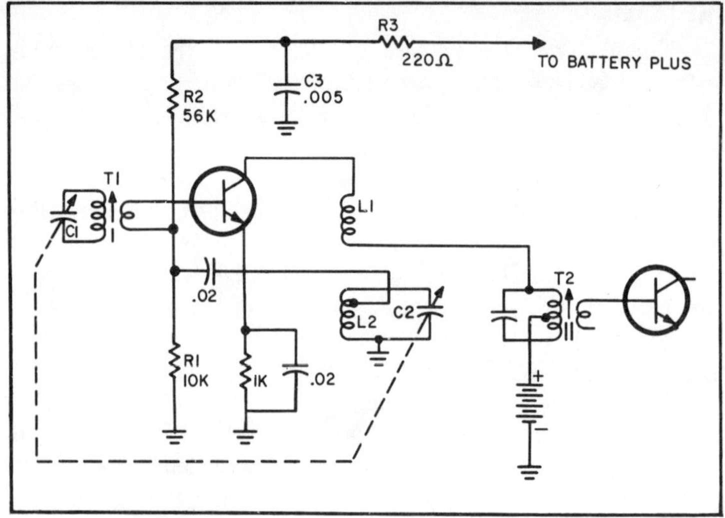

Fig. 6-21. Complete mixer circuit.

new components—C3 and R3. These form a decoupling filter to isolate the mixer from the other stages in the receiver.

HOW IT ALL WORKS

Our ride from the back end of a receiver to the front end is just about completed. At this time, though, let's think back over our whole trip to see if we can connect our ideas into one neat, little package.

The ferrite antenna in the receiver is capable of picking up any signal in the broadcast band. The desired signal, though, is selected by a variable capacitor shunted across the ferrite antenna. The broadcast signal, consisting of a radio frequency carrier modulated by an audio signal, is brought into the input of a mixer transistor. Here the signal is mixed or heterodyned with a locally produced signal. The result of this mixing process is a signal which represents the difference between the rf and local oscillator signals. This is a frequency conversion process and produces the intermediate frequency or i-f signal. All that has really happened is that our radio frequency has been lowered in frequency. It is still carrying with it, the audio signal from the broadcast station. The i-f is amplified by one or more i-f stages. Finally, when the i-f signal is strong enough, it is fed into a detector circuit. In the detector, the audio signal is rectified by means of a diode or a transistor. Immediately after rectification, the remaining i-f is shunted away or bypassed by means of a capacitor. The audio is fed into a driver stage, and then into a single-ended or push-pull audio amplifier, and finally to the speaker. Meanwhile, back at the detector, the rectified audio signal is also filtered and sent back to one or more earlier stages as an agc voltage.

Although each transistor stage does a slightly different job, you can see that they are similar in many respects. If you know how any transistor amplifier works, you are well on your way to understanding them all.

For this reason, the transistor rf amplifier shown in Fig. 6-22, will come as no great surprise. In this circuit, T1 now becomes an rf transformer coupling the rf stage to the mixer. The rf amplifier supplies the receiver with certain advantages. There is greater gain because of the amplification supplied by the rf transistor. There is also some added selectivity, since we now have three variable tuned circuits. This calls for a three-gang tuning capacitor on which we have mounted C, C1, and C2. Transformer T now becomes our ferrite rod antenna. The rf amplifier is neutralized by resistor R and capacitor C.

SSB AND FM RECEIVERS

So far in this chapter, we've discussed transistor detectors, agc, i-f amplifiers (which are actually just low-level rf amplifiers of a special sort), mixers and local oscillators. In fact, we've seen the ins and outs of a complete AM transistor radio receiver. But there are

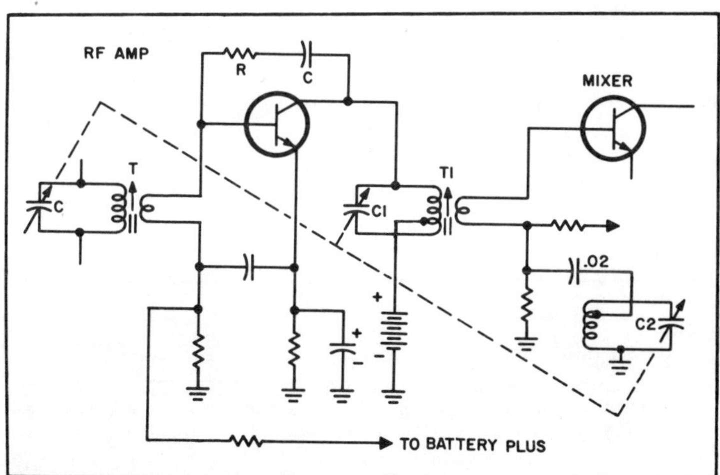

Fig. 6-22. Rf amplifier circuit. It can be connected to the mixer shown in Fig. 6-21.

other kinds of modulation besides AM, and these modes require different kinds of detectors. The simple detector/amplifier circuits in Figs. 6-1 and 6-2 are called envelope detectors, since they take the audio information from the signal envelope—the pattern we would see if we view the signal on an oscilloscope.

A single-sideband (SSB) signal needs a much different kind of detector, called a product detector. A frequency-modulated (FM) signal may be detected using either a discriminator or a ratio detector. We will now look at these kinds of circuits.

THE PRODUCT DETECTOR

An AM signal has certain frequency characteristics. Figure 6-23A shows how an AM signal looks on a plot of amplitude versus frequency, as we would see it on a spectrum analyzer, a lab instrument for examining the frequency characteristics of signals. In the case of Fig. 6-23A, the carrier frequency is 1,000 kHz and the sideband frequencies go about 5 kHz either side of the carrier. (The extent of the sidebands allowed depends on how far away, in terms of frequency, the nearest channel is. In AM broadcasting, the nearest signals are centered at 990 and 1,010 kHz; thus any bandwidth greater than plus or minus 5 kHz would create a conflict of interest.)

Actually, the audio information—all of it—is carried in the

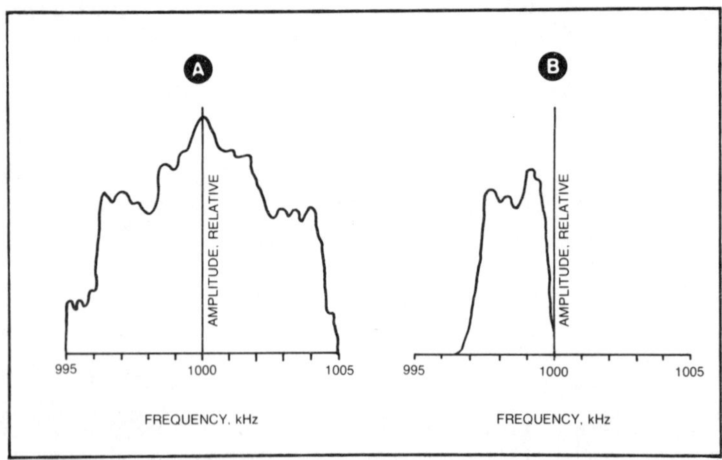

Fig. 6-23. At A, the frequency-versus-amplitude distribution for a typical AM signal. At B, an SSB signal. The example shown at B represents the lower sideband since it contains the sideband that is lower in frequency than the carrier. SSB signals are usually limited to 3 kHz above or below the carrier frequency, by means of special filters.

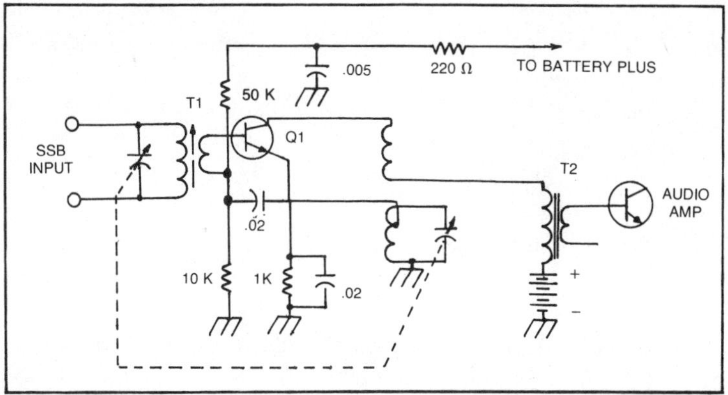

Fig. 6-24. A product detector. This circuit is essentially the same as a mixer, except that the output frequencies are in the audio range.

sidebands. And since the sidebands are essentially identical, one sideband contains all the information we need to understand what is being conveyed by the signal. The so-called carrier wave is not really necessary; it just goes along for the ride. Suppose we were to take the carrier out of the transmitted signal, leaving only the sidebands. The sidebands are rf; they are perfectly capable of propagating through space without the help of a carrier wave. This bit of trickery—removing the carrier from the transmitted signal—is pretty easy to do at the transmitter end of the circuit. It is also quite simple to replace the carrier again at the receiver end, using a local oscillator, and thus to get our original AM signal back.

Why would we want to take something away from a signal, only to put it back again? We don't, exactly. We go still further and snatch away one sideband at the transmitted signal end. This gives us a frequency distribution similar to that shown in Fig. 6-23B. A single sideband is all we need. It has no redundancy. It takes only half the spectrum space of an AM signal. The result is a much better signal-to-noise ratio for a given amount of signal power.

Our problem, then, is simply to replace the missing carrier at the receiver, on exactly the right frequency, so that we get an AM signal (with one sideband, but that's enough) out of the detector. Figure 6-24 shows how we might do this. You'll recognize this circuit as a close relative (almost an identical twin) to a mixer. What we're doing is beating the sideband frequencies of the signal together with a local oscillator, which is tunable so that it can be set to exactly the frequency the carrier would be if the transmitter were putting out an AM signal. The resulting audio frequencies may then be amplified in just the same way as the output signals from the envelope detector in an AM receiver. As a matter of fact, in our SSB receiver, we simply replace the envelope detector with a product detector. It's so simple that we may actually switch between one type of receiver and the other, just by changing this one stage. A double-pole, double-throw toggle switch will suffice!

THE FM DISCRIMINATOR

In contrast to AM and SSB signals, an FM signal maintains a constant amplitude at all times. The envelope and product detectors are of no value at all in deciphering the information from this kind of signal. What we need is a circuit in which a change in frequency will produce a change in the audio output. Such a circuit is shown in Fig. 6-25A.

When the FM signal is at zero frequency deviation—the carrier is centered—the discriminator has no output. But when the carrier frequency moves upward, the diodes rectify the output to produce a negative current. Likewise, if the carrier frequency decreases, a positive current is produced. Actually, the transformer is primarily responsible for the detection of the FM, since it converts the FM signal to an AM signal. The diodes simply act as an envelope detector for the resulting AM signal. This circuit has no amplifica-

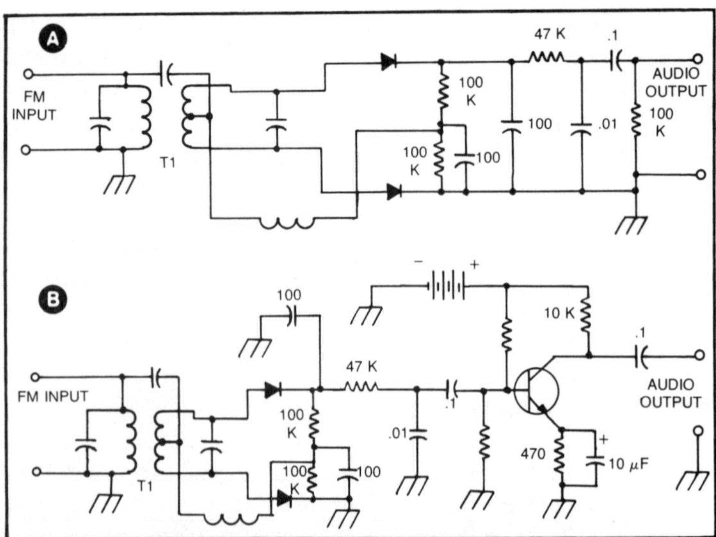

Fig. 6-25. At A, a discriminator circuit using only diodes. This circuit has no gain. At B, a transistor amplifier is added. The transformer T1 serves to convert the FM to AM by means of phase relationships between the primary and secondary.

tion, since it uses diodes and not transistors. But it is a simple matter to use a transistor also, as shown at Fig. 6-25B, if the number of stages must be kept to a minimum. Then, amplification as well as detection are provided by the circuit.

We must be careful when designing a discriminator circuit, to be certain that the relationship between carrier swing and audio output is linear. If it isn't, there will be distortion. The bandwidth of the FM signal being received must not be too great, or the discriminator will act in a nonlinear way. Some transmitters of FM have too much carrier swing; this is called overdeviation. Overdeviation will cause distortion even in a properly designed FM receiver.

The discriminator is sensitive to changes in signal strength, to a certain extent, as well as to changes in the frequency. In FM, we want our receiver to respond only to a frequency change. To make an FM receiver work ideally, then, we must put a stage just before the discriminator to keep the signal level as constant as possible. If there is an AM signal nearby, or if there is a thunderstorm or other source of electrical noise, the FM radio receiver should not be affected. We call such a stage a limiter because it "chops off" all signals at the same level.

How would we make a limiter circuit? One way is to put a transistor amplifier just before the discriminator, and bias it in such

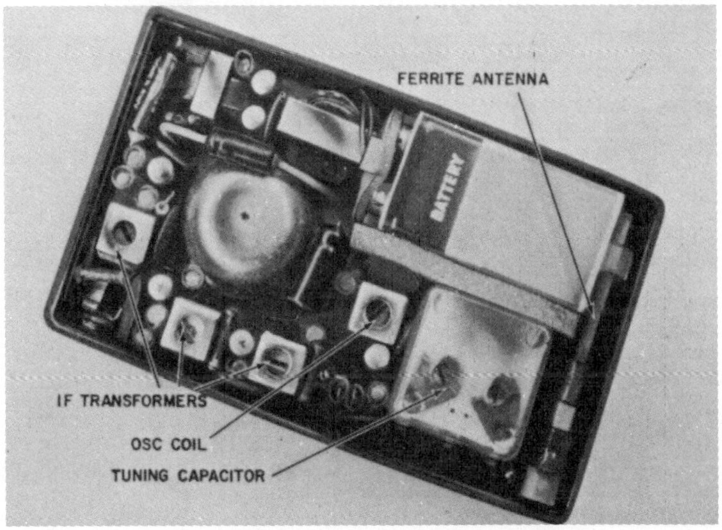

Typical transistor portable has three fixed tuned circuits in the i-f transformers and the tuning capacitor controls the resonant frequency of the ferrite antenna and oscillator.

a way that it is in saturation. This may seem strange! We are deliberately biasing a transistor improperly! But it serves a good purpose. Any further increase in the signal strength at the input of the limiter will cause no change in the output. But, of course, frequency changes will pass through with nary a hitch.

The limiter and discriminator form one kind of FM detection circuit. Another kind of detector for FM, called the ratio detector, acts as its own limiter.

MORE CIRCUITS

The circuits we studied in this chapter simply prove that transistors can perform the same functions as their predecessors, the tubes, in a receiver. But transistors certainly aren't limited to receivers. Transistors have taken the place of vacuum tubes in almost every application. We will now see how transistors can work in a transmitter.

QUESTIONS

1. What is the function of a detector?

2. What are the advantages of using a diode as a detector?

3. What are the advantages of using a triode transistor as a detector?

4. How must a transistor be operated to make it work as a detector?

5. What is automatic gain control? How does it improve reception?

6. How does an automatic gain-control circuit work?

7. What is the difference between ordinary agc and auxiliary agc?

8. What is decoupling? Why is it needed? How is it obtained?

9. What is meant by impedance matching? How is it done? Why is it necessary?

10. How is a transistor i-f stage neutralized? Why is neutralization used?

11. What is negative feedback? Positive feedback? When and why are they used?

12. What is a converter? Why is it needed? How does it work?

13. Describe a typical local oscillator and explain its operation.

14. Draw the basic diagram for a Hartley oscillator and for an Armstrong oscillator.

15. Compare mixers and converters and explain how they differ.

16. How does an rf amplifier improve reception?

Chapter 7

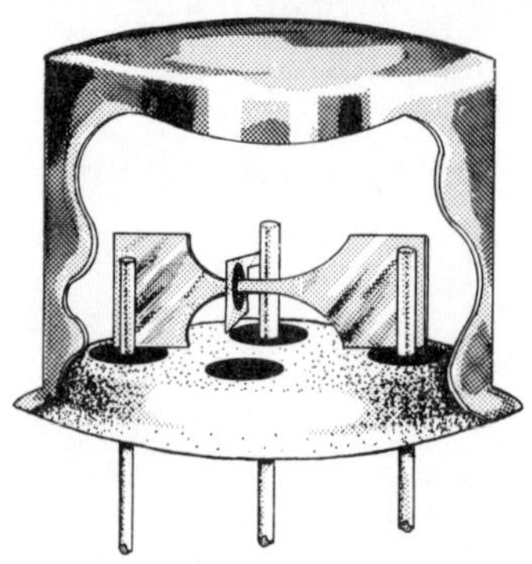

The Transmitter

NOW THAT WE HAVE SEEN THE WORKINGS OF A COMPLETE transistor receiver, let's turn things around and see how a transmitter operates. In many ways, a transmitter does things exactly backwards from a receiver, except that both are designed to amplify a weak signal and get a strong one. In the case of a receiver, we get high-power audio (enough to drive a loudspeaker, anyway!); with a transmitter, we get high-power rf energy. The actual amount of power may only be a few watts, or it may be thousands of watts.

CRYSTAL OSCILLATORS

In our last chapter, we saw how some simple types of oscillators work. We saw that an oscillator is pretty much the same thing as an amplifier, except that some of the output is fed back in phase to the input. This is done only at one frequency; otherwise, we would have little control over the frequencies at which our oscillator functions. We saw the Hartley and Armstrong types of variable-frequency oscillators (vfo's). We may use the same oscillators in a transmitter as we do in a receiver, but we should briefly look at one more kind of oscillator, designed to work on a fixed rather than adjustable frequency. This, of course, is the crystal-controlled oscillator.

Figure 7-1 is a block diagram of a simple continuous-wave (cw) transmitter. Notice that there are three oscillators—one vfo and

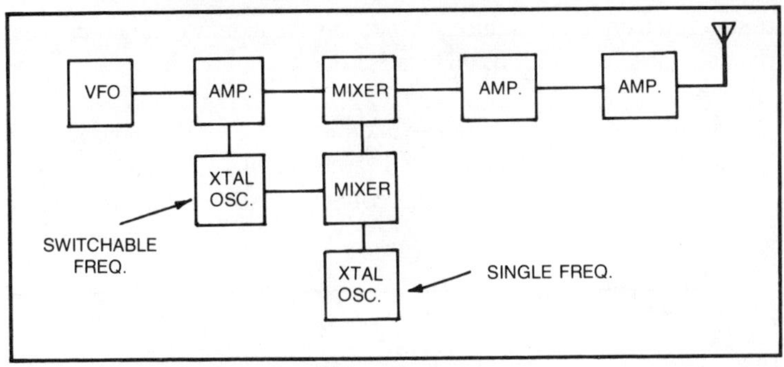

Fig. 7-1. Block diagram of a dual-conversion transmitter. We use this circuit so that we may design the vfo for a fixed range, say 4.5 to 5 MHz, and also to facilitate generating a multimode signal on a constant crystal-controlled frequency.

two crystal oscillators. Now why in the world do we need three oscillators to generate just one signal? Why can't we just use our vfo on the desired frequency? Well, of course, we can do it that way, and it's not too hard when our transmitter is required to generate a simple cw signal. But some transmitters have to function over a very wide range of frequencies. There are amateur transmitters, for example, capable of working all the way from 1.8 MHz to 29.7 MHz. It's pretty hard to design a vfo that will cover this frequency range effectively. So what do we do? We build our vfo to cover just one small frequency range—say, 4.5 to 5 MHz—and switch in different crystals in one of the crystal oscillator circuits. The other crystal oscillator is used for the sole purpose of generating a signal of the highest possible quality on only one frequency.

We can mix the two crystal-oscillator signals, and get some constant frequency that may be set according to the switchable oscillator. Mixing this with the output of the vfo, we can get any desired 1/2-MHz tuning range. For example, if we want to tune 7 to 7.5 MHz, we would use a 21-MHz crystal in the switchable crystal oscillator, and mix this with the output of a 9-MHz fixed oscillator. The resulting 12-MHz signal would yield the desired range when mixed with the vfo output: $12 - 5 = 7$, and $12 - 4.5 = 7.5$. In this case, increasing the vfo output frequency will cause a decrease in the actual transmitter frequency. You may have seen transmitters like this, where the vfo tunes one way for some bands and the other way for other bands.

Before the use of superheterodyne techniques in transmitters, frequency multipliers were needed to get the higher-frequency ranges using vfo's. This caused some problems, especially with respect to frequency stability, since any frequency change (drift) of the oscillator was multiplied by a factor equal to the harmonic order chosen!

Figure 7-2A is a schematic diagram of one of the most famous types of crystal oscillator, the Pierce oscillator. The crystal is connected between the base and collector. Notice that there is no tuned circuit in this oscillator, so the frequency is determined only

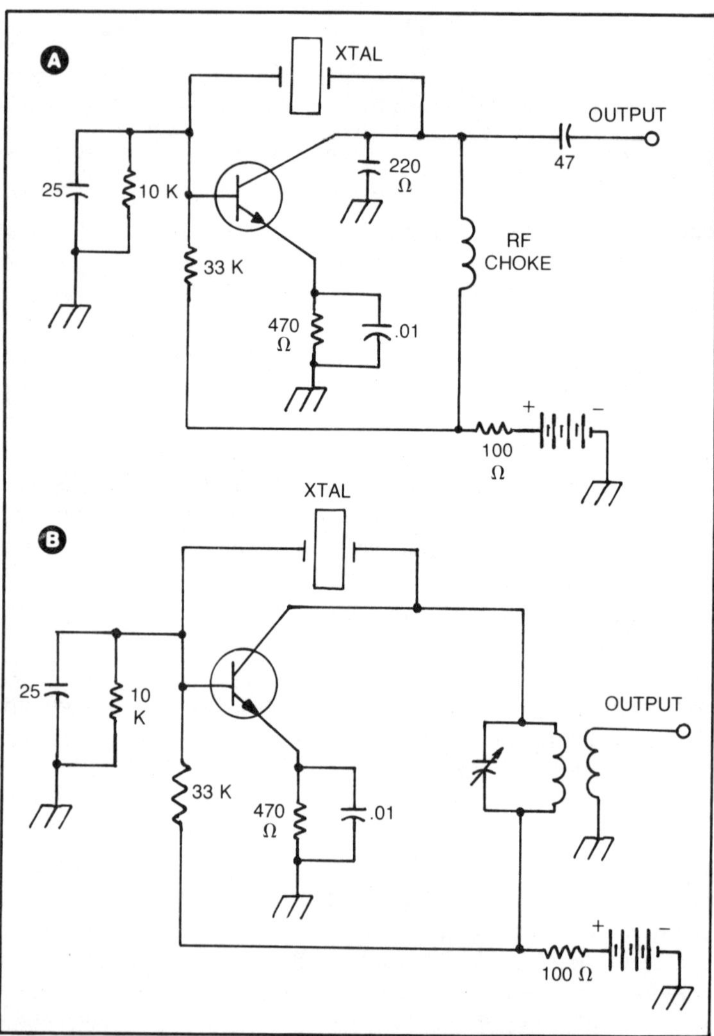

Fig. 7-2. At A, the Pierce crystal oscillator. The crystal is connected between the base and collector. At B, a Pierce oscillator using a tuned circuit at the output.

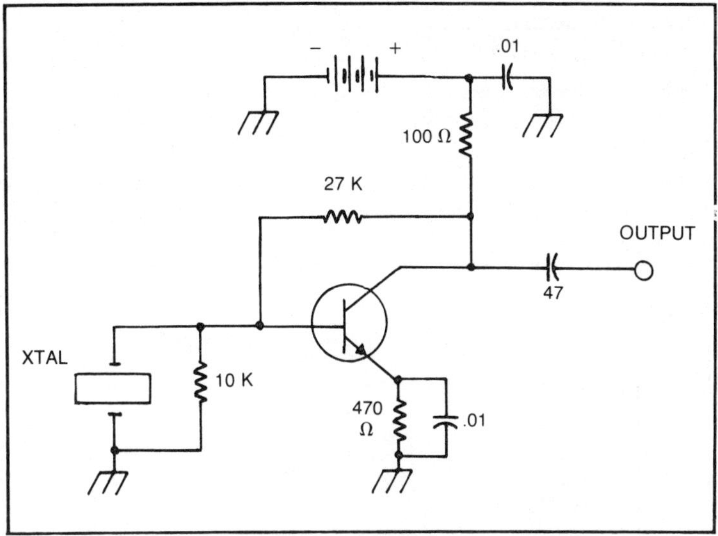

Fig. 7-3. Another type of crystal oscillator circuit.

by the crystal. Sometimes the output of this type of oscillator is rather "dirty," meaning that it contains frequencies at integral multiples of the crystal frequency. This may or may not present a problem. Harmonic energy may be suppressed by replacing the rf choke with a tuned circuit, as shown in Fig. 7-2B. The tuned circuit should be set to a frequency just a little higher than the crystal frequency.

Crystal oscillators are very stable. Their frequencies don't wander very much with changes in temperature or with mechanical vibration. By placing a small variable capacitor across the crystal, we can lower the oscillation frequency slightly, and if we choose a crystal a tad higher in frequency than we want, this allows precise adjustment of the frequency. This is especially important in ssb transmitters.

A somewhat different type of crystal oscillator is shown in Fig. 7-3. Here, the crystal is connected between the base and ground. This circuit acts as an amplifier as well as an oscillator. It is not a true Pierce oscillator, although you may sometimes see it called by that name. Just as with a true Pierce oscillator, harmonics may be reduced by replacing the rf choke with a tuned circuit.

We will examine the operational details of crystal oscillators more closely in Chapter 10. For now, though, we'll remain content with the knowledge that such circuits exist, and are of great advantage in transmitters.

THE AM MODULATOR

Oscillators by themselves were the very first transmitters; they were simply keyed on and off using Morse code. The Pierce oscillator just described could be used all by itself as a primitive low-power cw transmitter, of course restricted to the frequency of the crystal, by adding a key in the emitter circuit. But long ago, before transistors were even invented, somebody found out that

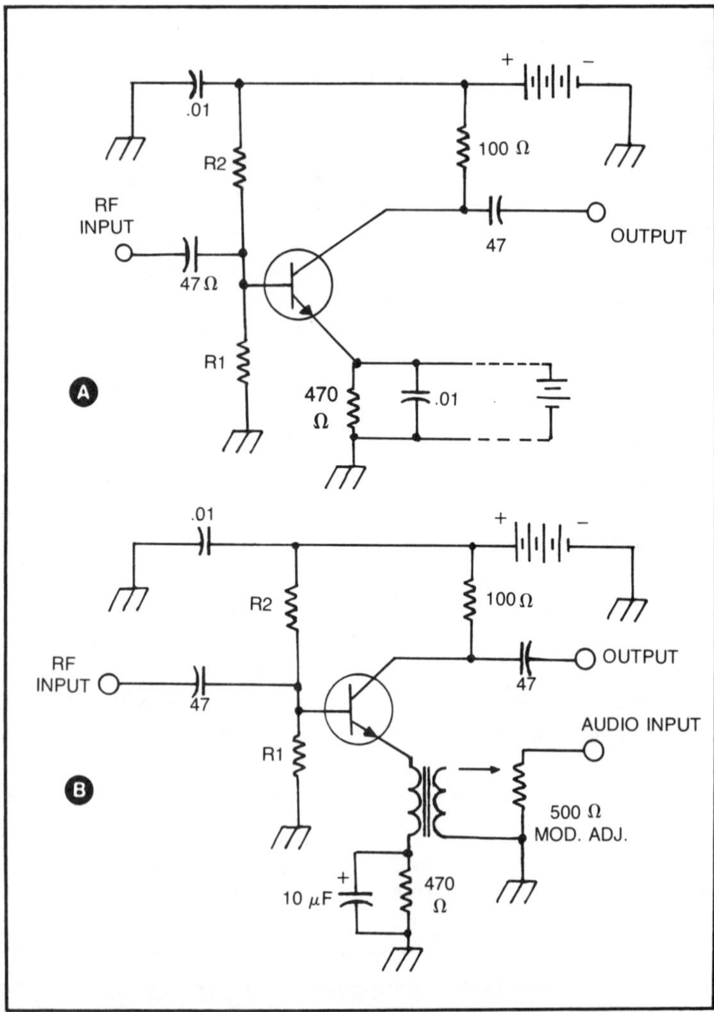

Fig. 7-4. An AM modulator is just an rf amplifier with variable gain. At A, a battery in the emitter circuit will change the gain of a properly biased circuit. Apply audio energy is similar in principle (B) to adding the battery.

voice information could be impressed onto an rf signal. Tubes were used in the first modulators, as these devices were called. But transistors are every bit as capable as tubes for this purpose.

An AM modulator is just a special sort of amplifier. Examine the circuit in Fig. 7-4A. You should recognize it as a simple rf amplifier of the common-emitter variety, using an npn transistor. There's nothing special about this amplifier; it's just an ordinary class-A circuit. The emitter is at rf ground, and the base potential is regulated for class-A operation by the voltage divider, R1 and R2. If we were to put a little battery across the emitter bypass capacitor, what do you think would happen? Yes, the gain of the amplifier would change. If the battery polarity were at negative ground, the gain would be reduced since the voltage between the base and emitter would be smaller. Similarly, if the battery were at positive ground, the gain would be increased (if the base bias were chosen just right) because of the increased voltage between the base and emitter. Actually, we want this circuit to be a class-AB amplifier—midway between class A and B, so that the gain will change either way depending on the polarity of the little emitter battery.

We have no reason, of course, to control the gain of this amplifier by putting a little battery in the emitter circuit. However, we may put a source of low-frequency alternating current there, such as an audio sine wave signal, as shown in Fig. 7-4B. We can adjust the level of the peak-to-peak audio voltage so that the transistor gain goes to zero when the signal reaches a positive peak.

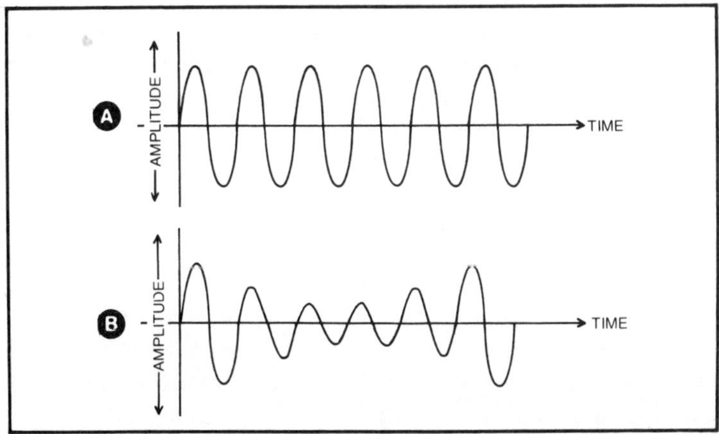

Fig. 7-5. The AM modulator has constant rf output when there is no audio signal, as shown at A. The audio signal causes fluctuation in the output (B).

The gain will increase somewhat at negative peaks of the audio signal. If we view the gain of this amplifier on an oscilloscope by looking at the output waveform, first without the audio signal applied (Fig. 7-5A) and then with it applied (Fig. 7-5B), we will see a very familiar sight: an amplitude-modulated rf signal! An AM modulator is really nothing more than a variable-gain amplifier.

Of course, Fig. 7-4B illustrates just one of many ways to obtain amplitude modulation. We call the method just seen by the name emitter modulation, for obvious reasons! But we could put the audio signal into the collector circuit, and then we would have (you guessed it!) collector modulation. If we apply the audio signal to the base circuit, we call it—well you have a good imagination. Anyway, the parallels with tube circuits are obvious if you have grown up with tubes. And if you haven't, don't worry.

THE SSB MODULATOR

In two-way communications, AM is hardly used anymore. You will still find it when you listen to short-wave broadcasts, and of course AM is still used in the broadcast band 535 to 1605 kHz. But AM makes inefficient use of spectrum space. We saw this in the previous chapter. Inefficiency is getting harder and harder to tolerate these days, when conservation seems to be the rule in all aspects of life. And it's no different in radio! So now we find single sideband (SSB) in widespread use for two-way circuits, especially in the short-wave bands.

As you recall, an SSB signal is just an AM signal without the carrier wave and one of the sidebands. To generate SSB, we must first employ a circuit known as a balanced modulator, which eliminates the carrier wave and leaves only the sidebands. Then we may use either a phasing circuit or a selective filter to choose either the upper sideband or lower sideband.

Figure 7-6 illustrates a simple balanced modulator. This circuit uses two transistors, with their outputs connected in push-pull. Remember the push-pull circuit for audio amplification, Fig. 5-13? Notice the similarity as far as the audio signal is concerned. But here, we're adding an rf carrier, so that each transistor performs the function of a modulator. In this case, we are using base modulation. But there's something a little strange, isn't there, about the way the rf signal is fed through this circuit! The two transistors are getting the rf in phase at their bases, but their collectors are connected to opposite ends of the output transformer. This causes the carrier wave to be canceled out. But the sidebands still remain in the output

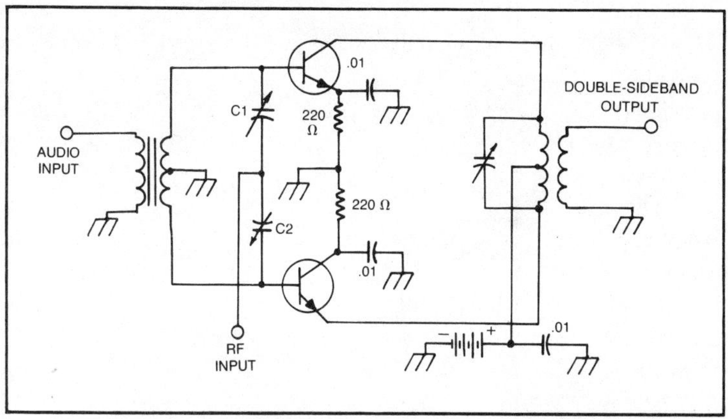

Fig. 7-6. A balanced modulator appears as a push-pull circuit for the audio energy, but the rf energy is canceled in the output since the two transistors receive rf input in phase.

of this modulator. Actually, the operation of this circuit is extremely critical. If one transistor has just the tiniest bit more gain than the other, the carrier will not be perfectly balanced out in the output transformer section. To make sure that the carrier is eliminated as much as possible, capacitors C1 and C2 are adjustable. With C1 set to approximately the middle of its tuning range, Q1 will get a certain amount of carrier input; ideally, C2 should be set to the middle of its range also, but perhaps Q2 has a little different gain than Q1 because of minute differences in the semiconductor material. Nobody's perfect! By adjusting C2, we can compensate for this possible difference, and get the greatest amount of carrier rejection by making the two signals equal at opposite ends of the output transformer.

The balanced modulator output is called a double sideband (DSB) signal, since it contains both sidebands but no carrier wave. To obtain a single sideband signal, we must use either a phasing circuit or a very carefully tuned crystal filter. Now it becomes apparent why we want to do this at a fixed frequency! Both phase and filter adjustment depend very much on the frequency of the suppressed carrier. How would we ever adjust a filter to exactly follow the signal of a vfo? It would be almost impossible! But with a crystal-oscillator input, it's easy; we can just "set and forget."

We may switch easily between AM and SSB in the same transmitter by employing the superheterodyne principle. We can also add FM without much difficulty; this, too, may be done at the crystal oscillator circuit if the required amount of deviation is not

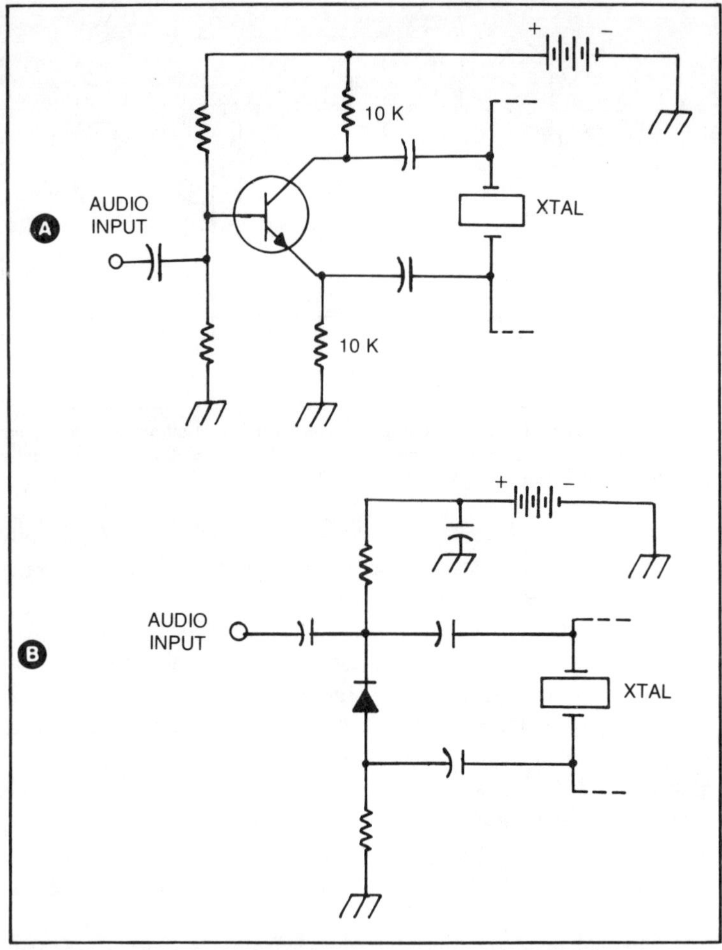

Fig. 7-7. Two methods of achieving FM with a crystal oscillator. At A, the transistor causes a changing capacitance to appear across the crystal. At B, a varactor diode does this with greater simplicity.

too great. Actually, frequency modulation is easier to obtain than AM or SSB.

FREQUENCY MODULATORS

A crystal may be "pulled" lower in frequency by the addition of capacitance across it. To get frequency modulation, we must find some way to rapidly vary a capacitance across a crystal or, if we are attempting to frequency modulate a vfo, across the tuned-circuit capacitor. Figure 7-7A shows how we might do this using a transistor. Ordinarily, the transistor conducts as a class-A amplifier, so that under conditions of no modulation, the crystal sees a small constant capacitance in shunt with itself. With positive audio signal peaks, the npn transistor conducts more effectively, and the crystal sees more capacitance. With negative modulation peaks, the transistor conducts to a lesser extent and the shunt capacitance appears to decrease. This will cause the oscillation frequency of the crystal to first get lower, and then higher. There is a practical limit to the amount of frequency swing that can be obtained in this way with a crystal, and thus we frequently see this reactance modulation done with a vfo.

Actually, the circuit of Fig. 7-7A is old-fashioned. We have looked at it to illustrate the principle of reactance modulation, and how the versatile transistor can be used for this purpose. But nowadays, a simpler circuit can be made using a special kind of diode called a varactor. This sort of diode acts as a capacitor at rf, and its capacitance changes with the voltage applied across it. Figure 7-7B illustrates a reactance modulator circuit using a varactor diode.

One method of increasing the amount of frequency swing from a voltage-controlled oscillator is to multiply the frequency. All rf signals have harmonics. They're unwanted guests in the output of a transmitter, and we try as hard as we can to get rid of them there (without complete success—the nasty critters!). So, as you might guess, they'll jump at the chance to be amplified and used! Any kind of nonlinear circuit, such as a class-B or C amplifier, has an output that is just teeming with harmonics. By adjusting the output tuning circuit to one of the harmonic frequencies (Fig. 7-8) instead of to the fundamental frequency, we have a circuit called a frequency multiplier. When we multiply the frequency of an FM signal by, say, a factor of 5, we also multiply the frequency deviation by 5. This gives us a way to get more deviation from a crystal oscillator than would otherwise be possible.

Whatever mode of operation we choose—AM, SSB, FM or cw—it's very convenient to have it at a constant frequency. We can move it around as we wish; we may choose the general range by mixing with a crystal oscillator having a switchable selection of crystals, and then get the exact frequency by means of mixing with a vfo. The mixers we use for this purpose are no different than those we discussed in the previous chapter. Any kind of signal is perfectly reproduced by mixing; there is no distortion of the modulation provided the mixer circuit is working right.

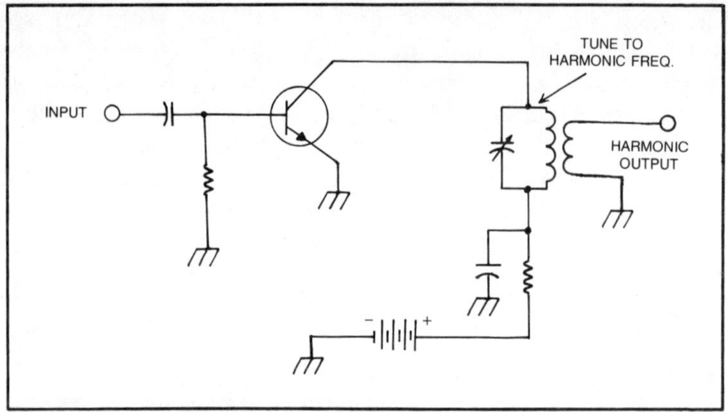

Fig. 7-8. A frequency multiplier increases the deviation of an FM signal, as well as increasing the frequency. The output may be heterodyned to any desired frequency without affecting the deviation.

Figure 7-9 shows a block diagram of the first few stages of a multimode (AM/SSB/FM/cw) transmitter. The output of the circuit in Fig. 7-9 is a low-power version of the signal we eventually want. The problem now is to amplify this signal, so that it has enough power to carry a long distance. This may be only one or two watts in some cases, perhaps 25 watts in others; and some transmitters are designed to deliver thousands of watts. The circuit characteristics depend on the mode and the frequency, as well as the amount of output power wanted.

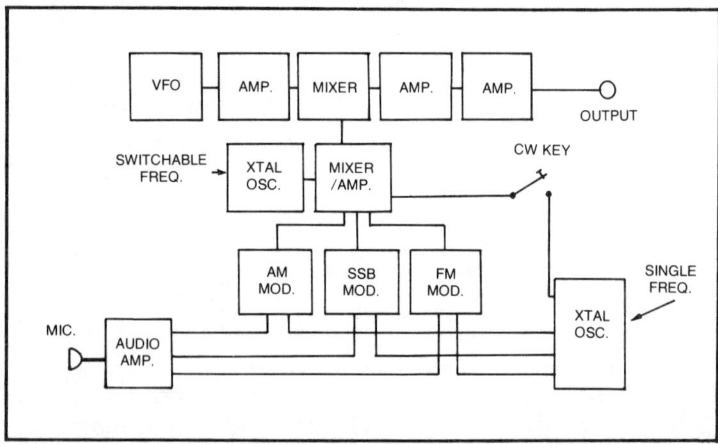

Fig. 7-9. A complete transmitter is shown here, except for the final rf power amplifier. This transmitter covers a wide range of frequencies and operates on cw, AM, FM and SSB.

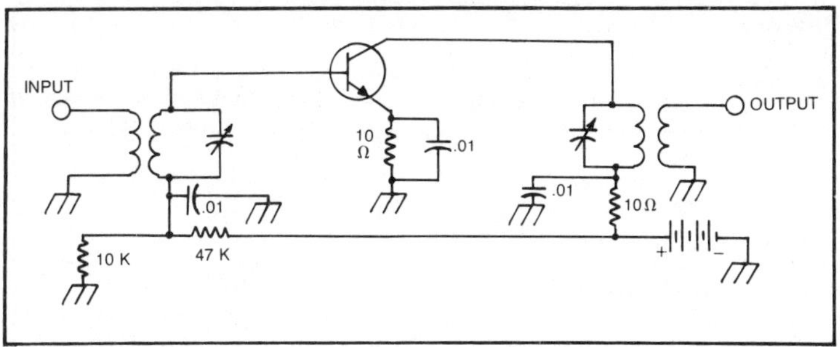

Fig. 7-10. A class A rf power amplifier is linear, but has rather low efficiency.

LINEAR AMPLIFIERS

For the next several pages, we're going to play engineer and see how various rf amplifiers work. We'll start with the circuit in Fig. 7-10, a straightforward class-A amplifier using a single transistor. This amplifier is pretty much the same as the class-A audio amplifier, except that there are tuned circuits in the input and output to discriminate against unwanted frequencies. (An rf signal has a very narrow bandwidth, percentage-wise, compared to an audio signal.) This amplifier, like all class-A amplifiers, has very little distortion. Thus it is a linear amplifier. But class-A amplifiers are notorious for their low efficiency, a bugaboo that we more or less have to put up with in audio work (unless we use class-B push-pull), but about which we need not suffer at rf. When amplifying rf energy, we have only to worry about the accurate reproduction of the modulation envelope. Whether or not the rf wave itself is perfect at the amplifier output is of no importance in most cases, since we can always filter out such distortion using tuned circuits. A class-B amplifier, called a single-ended amplifier since it uses just one transistor, will perform very nicely as an rf linear amplifier provided it is not overdriven. Figure 7-11 shows such a circuit. The only

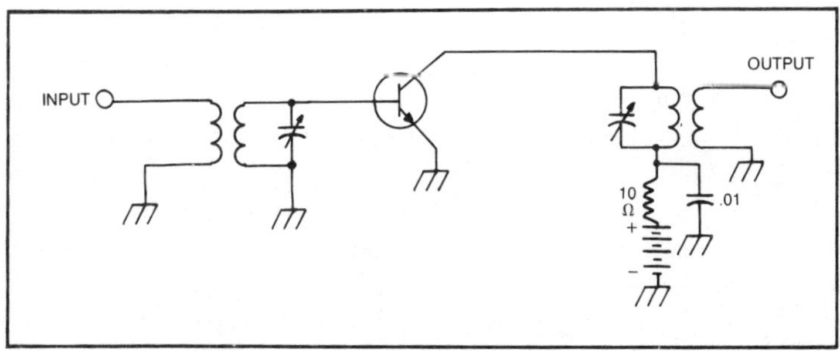

Fig. 7-11. A class B linear rf power amplifier.

difference between this and a class-A amplifier is the base biasing arrangement.

Now, wait! you may be saying. Why can we use a class-B circuit as a linear (and therefore supposedly nondistorting) amplifier at rf, but not at audio frequencies? Well, the answer is that we only need to be concerned with one thing: Is the output amplitude directly, and linearly, proportional to the input amplitude? The answer in this case is yes, even though half of the wave gets chopped off. We listen (in any receiver) to only the modulation on a signal, and don't hear any distortion of the wave itself. Class-B amplifiers are widely used as linear amplifiers. They are more efficient than their class-A cousins, even though class-B makes use of only half the cycle.

CLASS-C AMPLIFIERS

You might logically ask what is so sacred about linearity in a power amplifier. Actually, linearity is important only when the signal to be amplified varies in amplitude over some continuous range. While AM and SSB are like this, FM and cw are not. The amplitude of an FM signal is always the same; it's the frequency of the carrier wave that changes. A cw signal is either on or off, and there is no in-between condition. So we often see class-C amplifiers used for cw and FM.

To bias a transistor for class-C operation, we put the emitter-base junction well past cutoff by applying a reverse voltage. We very briefly mentioned class-C amplification back in Chapter 5. As you might guess, a class-C amplifier would sound just horrible for audio work, even if it were used in a push-pull arrangement, since collector current flows for less than half the time. You recall that in class-B, the signal will forward bias the emitter-base junction during half the cycle, and thus there will be collector current then. But in class-C, the input signal has to overcome the existing reverse bias, and collector current therefore flows only during the extreme peak of each cycle (Fig. 7-12)—perhaps 25 percent of the time. A class-C amplifier needs a pretty strong signal to begin with. You couldn't, for example, use a class-C amplifier as the first stage in a receiver. You would get no output at all! But the rewards of using class C are well worth the sacrifices! Class-C amplifiers have the greatest efficiency—ratio of power input to power output—of any kind of amplifier. In some cases it exceeds 80 percent.

Since class-C amplifiers are nonlinear, they create a great deal of harmonic distortion of their outputs. It is essential to have a

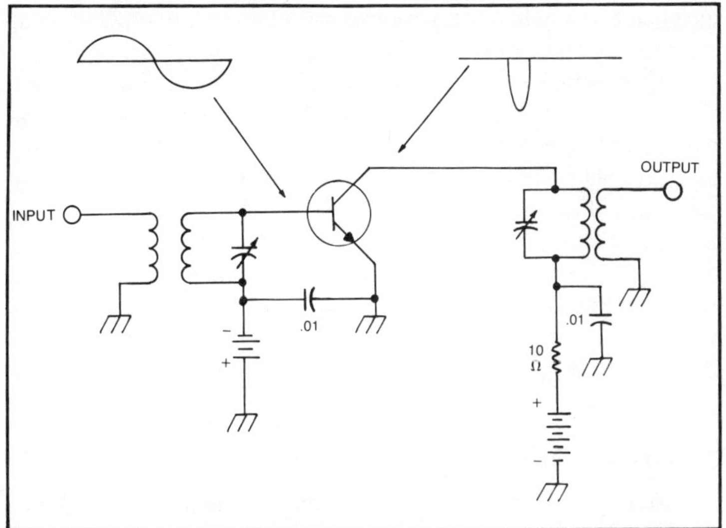

Fig. 7-12. A class C power amplifier has the greatest efficiency of any amplifier. It is not linear. Collector current flows only for a small part of the cycle.

highly selective tuned circuit in the collector arrangement of such an amplifier, and possibly even a low-pass filter as well.

Why is efficiency so important? Is the difference between a 70-percent-efficient class-B amplifier and an 80-percent-efficient class-C amplifier that overwhelming? Maybe not when the power output is small, say a half watt. But with 200 watts input, the difference would be 20 watts. Of what? Heat, of course—the heat that can destroy a semiconductor. Class-C operation lets us get the most out of a given transistor. The same transistor that might strain in class-B can take it easy in class-C.

CLASS-C MODULATORS

Putting an AM or SSB signal through a class-C amplifier would result in disaster. With AM, the negative modulation peaks (where the signal is at minimum amplitude) would be cut down to zero amplitude abruptly. With SSB, the low-level signal would disappear. But there is one technique that can be used with a class-C power amplifier to get AM. We apply a highly amplified audio signal right to the collector or emitter circuit of a class-C cw amplifier. This is known as high-level modulation, and an example of it is shown in Fig. 7-13.

The push-pull audio amplifier, Q1 and Q2, deliver their power into a special modulation transformer. The audio power output of the push-pull amplifier must be one-third the power output of the rf amplifier in order to have full (100 percent) modulation.

This technique may or may not be economical. True, we get more efficiency in the final rf amplifier than we do when using class A or B; for a given set of transistors, we can thus get the most power this way. But we have to develop a second source of high power, the audio amplifier; to get all this power requires several stages of amplification, each stage carefully engineered to minimize distortion. It's not easy to do. The modulation transformer is no little animal, either, if the transmitter must have a high output power level!

Of course, with SSB we have no choice. We cannot use high-level modulation, so we must use linear amplifiers.

PUSH-PULL CLASS B

You remember, of course, the audio push-pull amplifier. Well, this low-distortion method of getting lots of power can be used at rf as well. A properly operating class-B amplifier is linear as far as the modulation envelope goes, but it's a different tune (as we have seen) for the rf wave itself. Half of it is just cut off, and this creates a lot of harmonic energy.

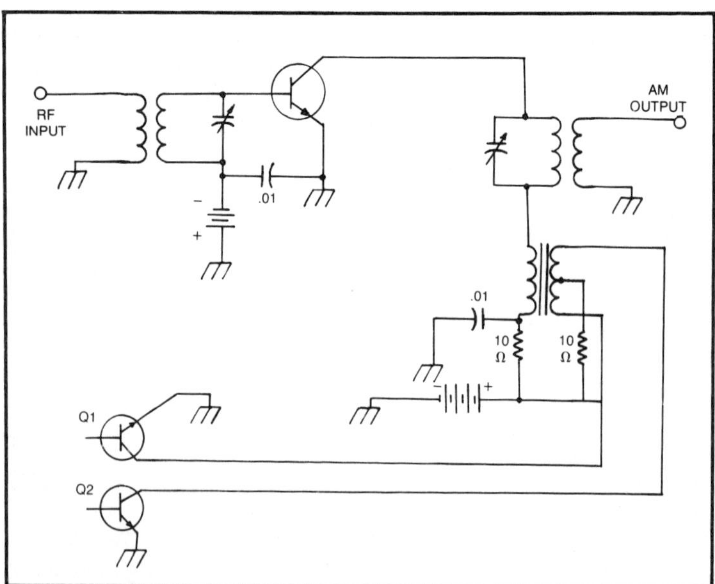

Fig. 7-13. Collector modulation of a class C rf power amplifier may be used to obtain amplitude modulation even though the amplifier is nonlinear.

Figure 7-14 shows an rf push-pull power amplifier. During one-half of the cycle, Q1 conducts and Q2 is cut off. During the opposite half of the cycle, Q2 conducts and Q1 does not. We get a nice, smooth alternating rf current in the output transformer.

The main advantage of push-pull amplification at rf is the excellent attenuation of the even-ordered harmonics: the second, fourth, sixth, eighth, and so on. These harmonics are effectively canceled out in the output circuit of a push-pull amplifier.

Actually, we do not see push-pull amplifiers used much in rf work nowadays. Part of the reason for this is the difficulty of obtaining precise balance between the two transistors over a wide range of frequencies. We might, for example, have two transistors that display exactly the same amount of gain at 1 MHz, but they will probably not be exactly alike at 15 MHz! This kind of ruins the advantage of push-pull operation, since unbalance means distortion, and distortion means harmonics. Another related phenomenon is called current hogging, where the two transistors show slightly different input impedances. The one with the lower impedance will end up holding the bag while the other transistor loafs along. This, again, results in a lopsided waveform, and if the proper precautions for current limiting are not taken, it can end with the destruction of one of the transistors.

Generally, tuned output circuits are used in all rf amplifiers except certain broadband types. Thus, harmonic distortion is not of great concern. We can just use a good, low-loss output coil and capacitor to maximize the selectivity.

Harmonics are not the only problem, though, that rf amplifiers face. There are other diseases too. One of these is the parasitic!

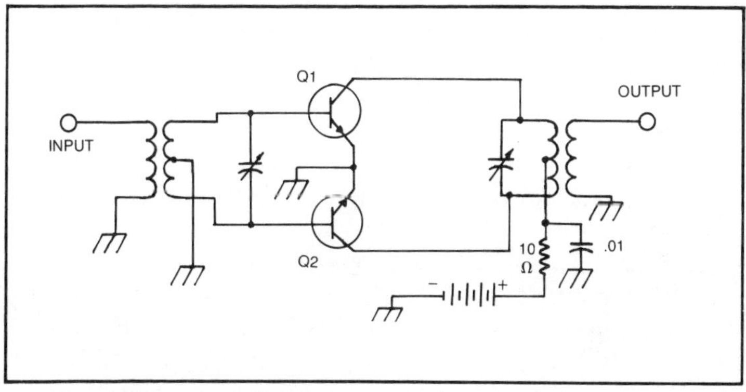

Fig. 7-14. An rf push-pull class B power amplifier. This is a linear amplifier. It tends to cancel even harmonics.

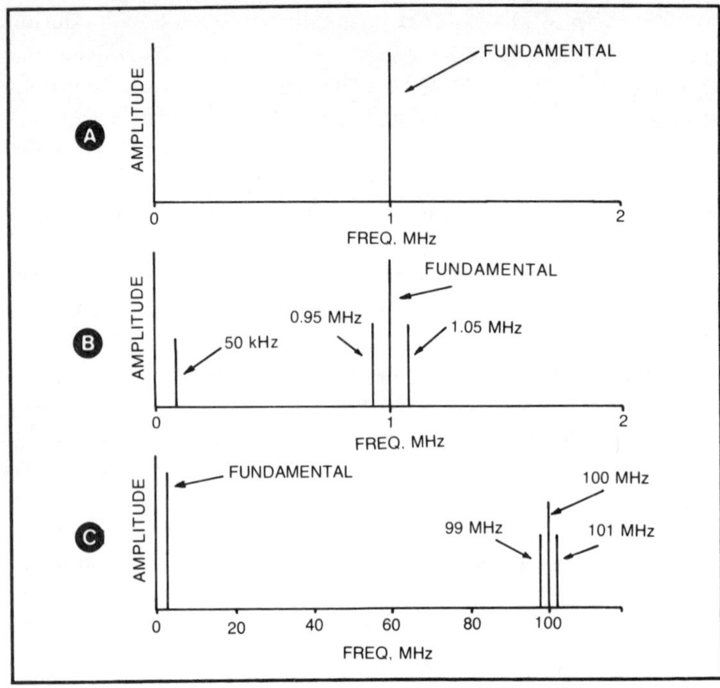

Fig. 7-15. At A, a spectral diagram of the output of a properly operating rf power amplifier at 1 MHz. At B, the result of 50-kHz parasitic oscillation. At C, the result of a parasitic at vhf (100 MHz).

PARASITIC OSCILLATION

Wouldn't it be nice if rf amplifiers would do their jobs without having to be held in check? Well, they do, sometimes. But not always. Sometimes they'll get the wrong idea and begin to oscillate. And it is usually on an entirely unpredictable frequency! This not only robs the amplifier of output at the desired frequency, but it can result in the transmission of many duplicate signals on unwanted frequencies.

Transistors may oscillate at very low or very high frequencies. Suppose an rf amplifier is operating at a frequency of 1 MHz with no parasitics, as shown in the spectrum illustration of Fig. 7-15A. Now what if a 50-kHz parasitic pops up—that is, the amplifier starts oscillating at 50 kHz as well as amplifying at 1 MHz? Then, sum and difference signals will appear at 1.05 MHz and 0.95 MHz, respectively. A 50-kHz signal will show up, too (Fig. 7-15B). The amplifier has suddenly become an oscillator/mixer/amplifier!

As if that isn't bad enough, parasitics usually are not very good signals. They aren't very frequency stable, and they often sound terrible, making our station seem haphazard (if anyone happens to hear our call letters on the wrong frequency).

In Fig. 7-15C, we see the result of a vhf parasitic at 100 MHz. Now, there are signals not only at 1 MHz, but at 99 MHz (100 – 1 MHz), 100 MHz, and 101 MHz (100 + 1 MHz). Harmonics of all these frequencies will probably also be present, as well as their intermodulation products, literally peppering the electromagnetic spectrum with replicas of our signal.

All right! Point made! How do we keep an rf power amplifier from being figuratively eaten alive?

Figure 7-16 shows one method of eliminating low-frequency parasitics in a solid-state rf power amplifier. The capacitor and resistor, connected in series between the collector and base, provide negative or degenerative feedback. The exact values of the resistor and capacitor depend on the operating frequency of the amplifier. (The parallel resistor and capacitor in the emitter circuit also provide some degeneration.) Since some of the collector signal, 180 degrees out of phase with the base signal, is fed back to the base, the addition of these components reduces the overall gain of the amplifier a little bit at the operating frequency. We can make up for this by increasing the drive power to the amplifier.

For vhf and uhf parasitics, we may place ferrite beads around the leads to the base and collector of the transistor (Fig. 7-17). These look like tiny doughnuts, made of powdered-iron material with high magnetic permeability. When these beads are inserted, they act as vhf/uhf chokes, but pass essentially all the energy at the operating frequencies, assuming the operating frequencies are relatively lower than vhf or uhf. Usually, ferrite beads need be used

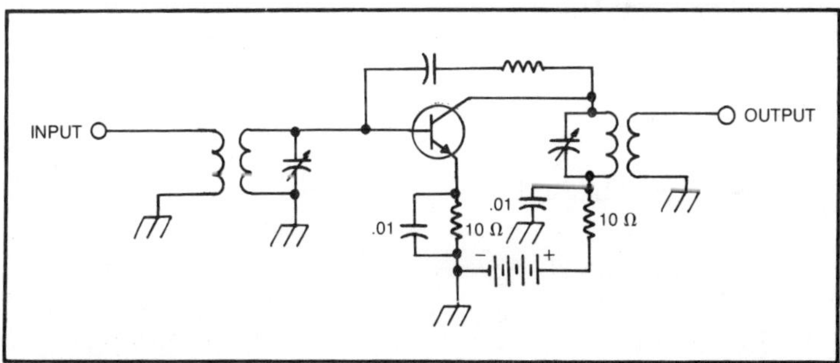

Fig. 7-16. Eliminating low-frequency parasitics is accomplished by placing a resistor and capacitor in series between the collector and base to provide negative feedback.

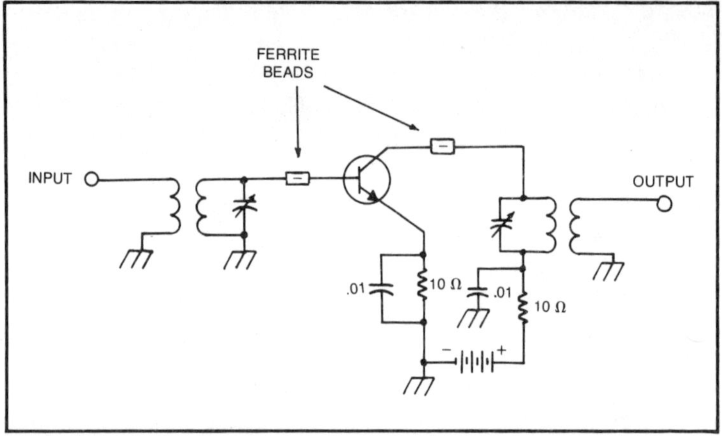

Fig. 7-17. To eliminate vhf and uhf parasitics, ferrite beads may be placed around the base and collector leads. This reduces the gain of the amplifier at higher frequencies, preventing accidental oscillation.

only around the base lead; they might be damaged in a high-power amplifier if placed around the collector lead.

BROADBAND POWER AMPLIFIERS

So far, all the rf power amplifier circuits we have seen have been of the tuned output variety. There are some good reasons for this! For one thing, a tuned output is itself a sort of parasitic choke, since it favors the operating frequency but casts a disapproving eye on all other frequencies. Second, tuned circuits increase amplifier gain and efficiency because they are resonant, and tend to concentrate rf energy at the desired frequency. Third, tuned circuits help to suppress harmonics, and also any other unwanted signals that may find their way to the amplifier input (for example, a tiny bit of the local oscillator or vfo signals). Still, tuned circuits require time and effort to adjust, and they must be reset every time the transmitter frequency is changed by more than a few kHz. The desire for time-saving convenience rules all aspects of life, it seems; and so it is not surprising that broadband power amplifiers have come into existence!

A broadband amplifier *must* employ some sort of low-pass filter at the output to discriminate against harmonic energy. There is no escaping this, since we have no other means of frequency selection at the output. Oh, yes, we still can (and should) employ some way to prevent parasitics at extremely low or high frequencies. But to keep harmonics down, as well as intermodulation feedthrough, a broadband circuit must have at least some output filtering. To ignore this would be to court disaster!

Figure 7-18 illustrates a class-B push-pull broadband linear amplifier. The input and output transformers are designed to be functional over a wide range of frequencies, with very little change in attenuation from the lowest to the highest desired frequency. The low-pass filter at the output of the amplifier should have a cutoff frequency about 1.5 times the operating frequency, but in practice may vary between about 1.1 and 1.7 times the fundamental. (Even though a push-pull circuit provides excellent second-harmonic attenuation, we may still need filtering at the output.) As the operating band is selected by switching crystals at the local oscillator, different components may be switched into the low-pass filter as well, using multiple switches on a single shaft.

We may employ class-C amplifiers as broadband devices, and we may also use class-A or class-B single-ended circuits. Some transistor types are especially designed for class C operation, while others are intended mainly for linear applications. Any of them can be used in broadband as well as tuned rf power amplifiers.

IC POWER AMPLIFIERS

Occasionally we will see integrated circuits designed for rf power amplification. Such circuits contain all the peripheral components needed for a one- or two-stage amplifier, as well as the actual transistors, packed inside a rectangular package. An example of such an IC is shown in the photograph at Fig. 7-19. This particular IC is capable of delivering up to 35 watts of rf power output continuously in the 140 MHz range. The heat is dissipated through the bottom of the IC, which is thermally joined to the chassis using a special silicone compound. This device is a class-C broadband

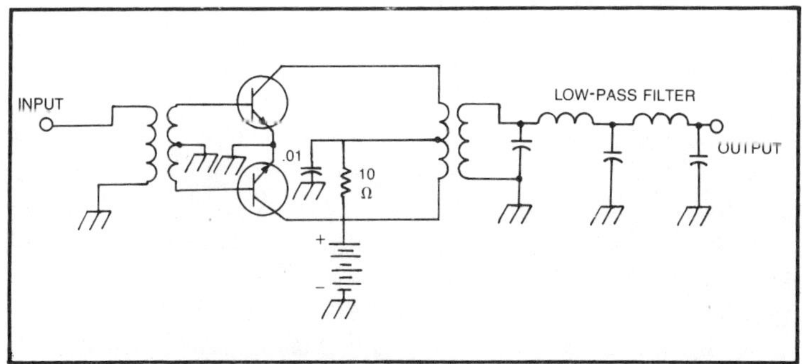

Fig. 7-18. A broadband push-pull rf power amplifier.

Fig. 7-19. Some rf power amplifiers are entirely contained in a single integrated-circuit package. (Photo courtesy International Electronic Systems, Inc., and Mark Thomas)

amplifier, and is used in the final stage of an amateur FM transceiver. Integrated circuits for rf power amplification are available in various power sizes, and for linear as well as class-C applications.

TRANSISTORS IN PARALLEL

One method of getting greater power output with transistor amplifiers is to connect several of them in parallel. Then, if the circuit is properly engineered, each transistor will dissipate a relatively small amount of power, while the total output is much greater. Figure 7-20 illustrates an example of a parallel-transistor rf power amplifier.

We have to be extremely careful in designing an amplifier such as this. There are a few potential pitfalls! Of greatest urgency is the prevention of current hogging. One transistor may get stuck with all the burden while the rest just loaf along, taking a free ride! This is especially apt to happen in a parallel-transistor amplifier. Of course, all the transistors should be the same type (that is, they should have identical manufacturers' part numbers, such as 2SC741 or whatever). But even then, one transistor will probably differ slightly from another. Emitter resistors are absolutely necessary! These resistors should be chosen to prevent excessive current from flowing through any given transistor, but if they are too large, the amplifier gain will suffer.

Another bugaboo of parallel-transistor amplifiers is their exceedingly low input impedance. A single-transistor rf power amplifier has a pretty low base impedance already; it may be from 3 to 5 ohms. When we put devices in parallel, we lower the input impedance by the same factor as the number of transistors: Two of them means half the impedance of either, three means one-third the base impedance of a single transistor. It can get difficult to provide a good impedance match to each individual transistor when their composite input impedance is so low. For example, we might use

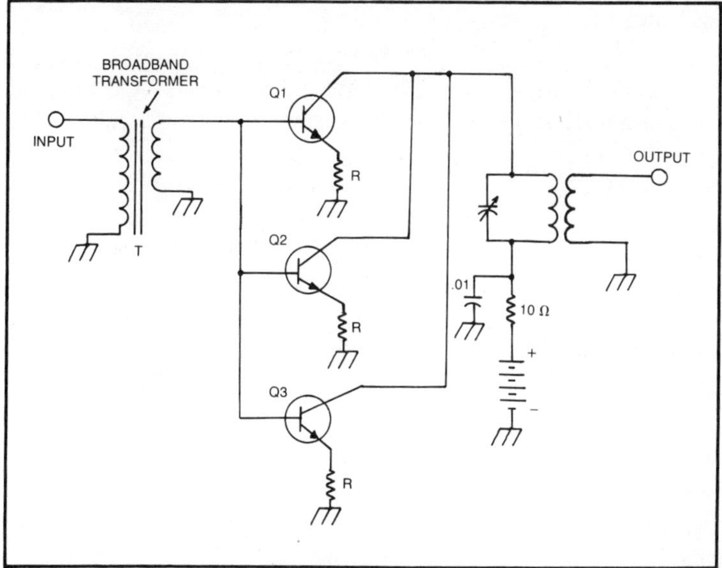

Fig. 7-20. Connecting transistors in parallel allows greater power output than would be possible with a single transistor. The emitter resistors (R) prevent current hogging. The transformer (T) assures the proper impedance matching to the low-resistance base circuit.

four transistors in parallel, each one with a 4-ohm base impedance. If the amplifier must show a 50-ohm input impedance, then, we'll need a 50-to-1 stepdown transformer at the input! Such transformers can be built, of course, but their turns ratios are critical and they are hard to design for wideband use.

What's all this about matching of impedances? Who cares? We discussed it briefly in the last chapter. In order to get good transfer of power from one stage of amplification to the next, we must match the output impedance of each stage to the input impedance of the next. To ignore this could mean a great loss of power, since one mismatch may be reflected back through several succeeding stages. This would reduce the drive power to the final amplifier to just a whisper of its intended self.

Impedance matching is usually done, in rf power amplifiers, by using broadband transformers. A transistor may also serve this function. When connected in a common-emitter or common-base configuration, we have a stepup transformer; in the common-collector mode with the output taken from the emitter circuit, we get a stepdown transformer. The precise matching ratio depends on the values chosen for the biasing resistors.

AND SO . . .

We have now seen how transistors can be used to build a complete receiver and a complete transmitter, for cw, AM, SSB and FM. As a matter of fact, transistors have just about taken over the role of tubes in all aspects of radiocommunications. The only exception is the very high-powered transmitters—many thousands or even millions of watts—where tubes still prevail because no transistor is rugged enough to dissipate all that heat! But maybe some day there will indeed be a megawatt transistor.

In our next chapter, we will investigate some other kinds of transistors besides the simple pnp and npn types. Transistors aren't limited to use in transmitters and receivers. We find them everywhere.

QUESTIONS

1. What is the purpose of a modulator?

2. What is the function of a balanced modulator? A reactance modulator?

3. What advantages are offered by use of superheterodyne circuits in transmitters?

4. What is high-level modulation?

5. Why can't we use a class-C circuit to amplify an SSB signal?

6. Name one advantage of push-pull operation in an rf amplifier.

7. What are parasitic oscillations?

8. How might we prevent low-frequency parasitics? Very high-frequency parasitics?

9. What is the main disadvantage of a broadband amplifier?

10. What is current hogging? How can it be prevented?

11. Why is impedance matching important between amplifier stages?

12. What are harmonics? How can they be minimized?

Chapter 8

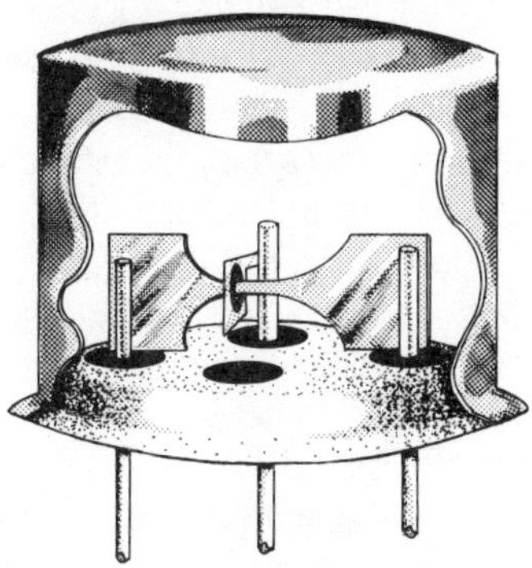

Transistor Types

THE TRANSISTORS WE HAVE BEEN DISCUSSING THUS FAR HAVE all been lumped into two very broad and general categories. They have been either pnp or npn units, but other than the fact that we have designated them as triodes, we have said nothing to you so far to give you a hint that there might be some very strong differences in transistors. The electronic symbols for an audio power transistor and a converter transistor will look exactly alike, and yet the operating requirements for these transistors are opposites. The output transistor must handle power at comparatively low frequencies. The mixer is much less concerned with power but works at considerably higher frequencies. Obviously, since their functions are so dissimilar, there must be considerable differences in the way they are fabricated and packaged.

TRANSIT TIME

Although we may be inclined to look on the electron as a speedy little cuss, yet electrons can be slowed when they encounter obstacles and barriers, just like people. Consider the electron in a vacuum tube. If we have a triode, nothing (with the exception of the control grid) stands in the way of a rapid passage from the cathode to the plate. And even here we make the grid as much of an open mesh as possible so as not to interfere with electron travel.

But what do we have in the transistor? Instead of open space, we have a crystal lattice structure, and so we are, in effect, forcing the electron to hop from one atom to the next. But the frequency at which we can operate is determined by how fast we can get the electron from the emitter to the collector or from the collector to the emitter. And, quite unlike the tube, we always have a barrier between these two electrodes. This barrier is the base, and while we can make the base extremely thin, the fact remains that the base exists between the collector and emitter. (A typical value of base thinness would be .001 inch or less.)

There are other factors which tend to limit the frequency range—that is, the upper limit of frequency at which the transistor will operate. One of these is the capacitance existing between the transistor electrodes. In Fig. 8-1A we see how we can come to regard the transistor as a capacitor. Here we have the pn section and, as you will observe, the diode is reversed biased. Current flows initially, but ceases shortly thereafter, because of the build-up of opposing charges at the junction between the two blocks of semiconductor material. But isn't this action very reminiscent of the way a capacitor charges? When we connect a capacitor to a battery as in Fig. 8-1B, there is a momentary rush to charge the capacitor. The current flowing in the circuit rapidly drops as opposing charges collect on the plates. In the transistor, the effect is just as though we had a capacitor in shunt with the elements of the transistor. But as we go up in frequency, the reactance of this shunting capacitor becomes less and less, and, if the frequency is high enough, acts like a virtual short across the transistor.

There is still another factor with which we must contend when considering transit time and that is the material of which the base is made. We will get swifter movement of electrons through certain semiconductor materials than through others.

ALPHA CUTOFF FREQUENCY

As we increase the frequency of a signal to be amplified by a transistor, there will come a point at which we will simply be wasting our time, for there will be no amplification. The transition from a condition of gain to one of no-gain isn't an abrupt change. As a basis for comparison among all transistors, we can establish the gain of a transistor at 1 kHz as standard. We can consider the practical operating limit to exist when the gain drops to .707 of the gain at 1 kHz (as we increase the frequency). This is known as the alpha cutoff frequency.

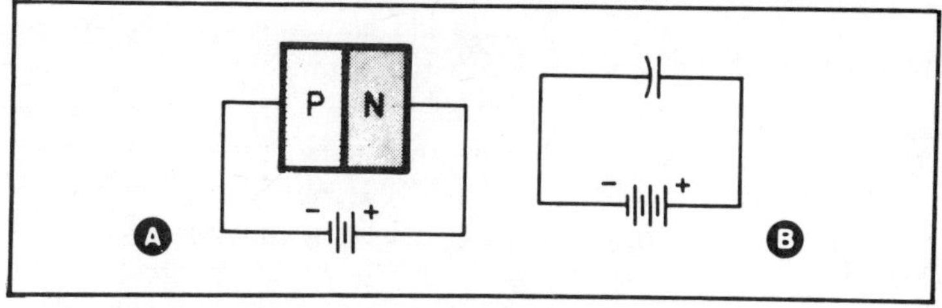

Fig. 8-1. The diode has some resemblance to a capacitor.

THE TETRODE TRANSISTOR

We have now revealed the base as one of the chief culprits in limiting alpha cutoff frequency. This means that if we want to increase the operating frequency of a transistor we must force the base to yield to some sort of treatment. There are various pleasant ways of doing this, one of which results in a unit known as a tetrode transistor.

At this point you have probably jumped headlong to a wrong conclusion, based on your experience with vacuum tubes. In the tube we make a tetrode by adding a fourth element known as a screen grid. The name tetrode means four elements, so applying it to the transistor is somewhat misleading.

In Fig. 8-2 we have a block diagram of a tetrode transistor. All

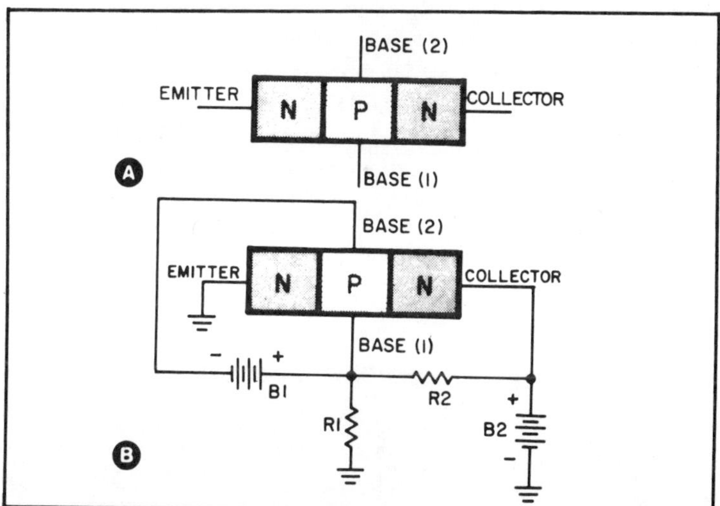

Fig. 8-2. The tetrode transistor gets the effect of a reduced base area by applying an additional bias to it.

we have really done, though, is to add another lead to the existing base. Basically, what we have is a transistor, with the added lead biased negatively with respect to the base. The transistor can be either a pnp or an npn type, but, as you can see, the one in Fig. 8-2 is an npn. Since we now have two base leads, one is identified as base 1 and the other as base 2. The transistor is forward and reverse biased in the usual way. R1 and R2 are the biasing resistors for the base-emitter input circuit. A single battery, B2, supplies both forward and reverse bias for the transistor. However, we have an additional battery, B1, connected between base leads 1 and 2.

Now examine the lead going to base 2. It is connected to the negative terminal of B1, and so, apparently, we have made the base negative with respect to the emitter. But this isn't entirely correct, since the other base lead (#1) is connected tapped up on voltage divider R1-R2. But how do we want the base biased? Since we have an npn unit, we want the base to be positive with respect to the emitter. The effect of B1, then, is to reduce the amount of area of the base that is really positive with respect to the emitter. But reducing the area gives us the same effect as reducing the volume of the base. Since the alpha cutoff frequency depends on the thickness of the base we have increased our available operating frequency. Actually, the situation is a little better than we expect, since the alpha cutoff frequency varies as the square of the thickness of the base. Thus, if we manage to cut the thickness in half, we raise the alpha cutoff frequency by a factor of four. If we cut the thickness to one-third of the original, we increase the alpha cutoff frequency by nine. As you can see, the rewards are great for this effort.

Just as we have symbols for the npn and pnp triodes, so too do we have symbols for npn and pnp tetrode transistors. These are shown in Fig. 8-3.

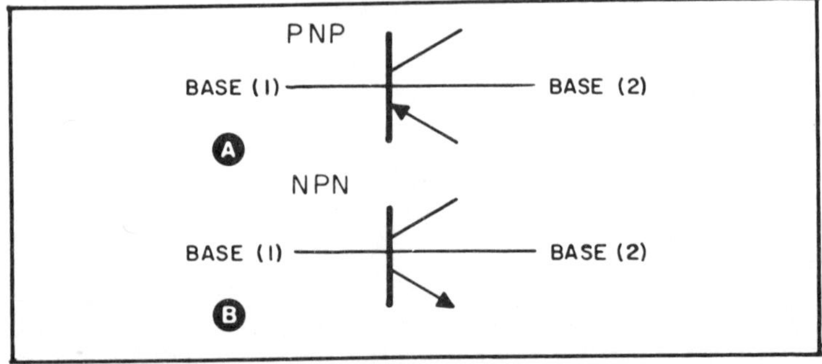

Fig. 8-3. Electronic symbols for pnp and npn tetrode transistors.

THE SURFACE-BARRIER TRANSISTOR

Since our objective (at the moment) is to get a high-frequency transistor and since we can do this by making the base thin, why not shave the base down as much as possible? This may sound terribly difficult, since the thickness of the base is often fractions of a thousandth of an inch, but it can be done. When we do this, we solve one problem only to be faced with another. How do we fasten a lead to something that must be so thin? A pleasant and practical solution to this problem is achieved by the surface-barrier transistor. We have a sliced-through section of one of these in Fig. 8-4.

Most of the base remains just as it is in an ordinary transistor. However, in the region between the emitter and the collector, the base material has been etched away, and as a result the emitter and the collector may be separated by as little as 0.01 inch. Since the remainder of the base is as thick as it was before, there is no additional problem of fastening the base lead.

We now have two techniques for reducing base thickness. One is electrically, as in the tetrode transistor, and the other is physically, as in the surface barrier. In both cases we must pay a price to get what we want, and the price is a reduction in power handling capabilities. An electric strain exists between emitter and collector due to the battery. By reducing the thickness of the base, we have really brought the negative and positive terminals of the battery much closer (electrically) to each other.

The surface barrier transistor, though, is not quite the same as all the other transistors we have studied. The base is a single crystal or wafer of n-type material and that is all—no p-type material is used. The emitter and collector electrodes are made of a metal called indium. These metals are plated onto the etched out part of the n-type material wafer. The metal does not melt or diffuse into the material, simply forming a plated electrode. We do have a barrier, though, just as in any ordinary transistor. This barrier exists between the plated electrodes (emitter and collector) and the base, and as a result we get the equivalent of transistor action.

The symbol for the surface-barrier transistor is the same as for a pnp unit, and is treated as such with respect to bias. That is, the emitter is made positive, the base less so, while the collector is negative with respect to the emitter.

The element, indium, selected for the emitter and collector of the surface-barrier transistor, belongs to the acceptor group, gallium, boron, and aluminum, mentioned in an earlier chapter. Since it has but three electrons in its outermost ring, indium is short

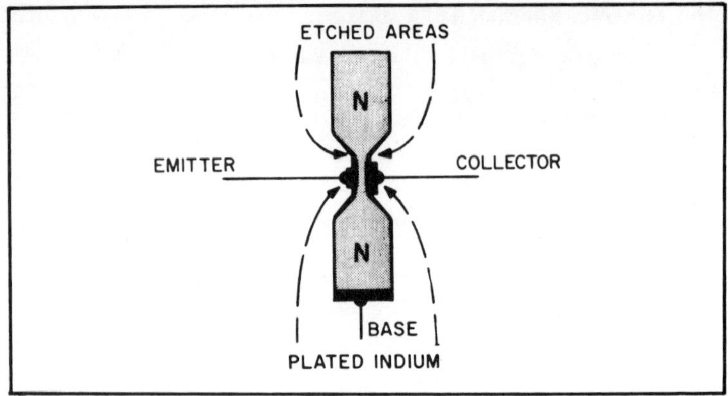

Fig. 8-4. The surface-barrier transistor uses a single section of n-type material. The effect is that of a pnp transistor.

an electron, hence the p-type designation. For this reason, all surface barrier transistors using indium for emitter and collector, and n-type material for the base must be of pnp type.

There is a very important difference between the surface-barrier transistor and all the others we have studied so far. In the usual transistor, the semiconductor material is doped with an impurity—either donor or acceptor type—to form n-type or p-type semiconductor material. In the surface barrier transistor, however, just one element is so treated—and that element is the base, made of n-type material. The other electrodes, the emitter and the collector, consists of a small plated area of a metal—indium.

THE DIFFUSED TRANSISTOR

There is still another type, the diffused transistor, which seems to resemble the surface-barrier transistor, yet possesses a very significant difference. The diffused transistor appears in Fig. 8-5. In this transistor, as in the surface-barrier transistor, we start with a section of n-type material. This is our base. A small amount of indium is melted into the base on each side, forming the emitter and collector. The melting process is continued until the emitter and collector almost touch each other, resulting in a very narrow base region between them. This is just what we want, of course, to permit a high alpha cutoff frequency.

When the indium melts into the material, we get the effect of doping the immediate region around the indium. And since indium, as we mentioned earlier, is an acceptor impurity, the result is that we get the formation of p-type material in the region of the impurity.

Note the difference between the surface barrier and diffused pnp transistors. In the surface-barrier transistor, the indium is just a plated electrode. There is no physical penetration into the n-type base. The surface-barrier transistor depends on an etching process to reduce the base thickness. The diffused transistor depends on the melting of the indium to narrow the base. For both types of transistors, the symbol is identical—and is that of a pnp transistor.

MICRO-ALLOY TRANSISTOR

There are different methods used in putting the emitter and collector electrodes onto a surface-barrier type transistor. They can be plated on or they can be alloyed as a very thin film of metal on the n-type material base. The produces a unit known as a micro-alloy transistor. An alloy consists of two substances that are melted or diffused into each other and so, in this respect, the micro-alloy transistor is more closely related to the diffused transistor.

In working with transistors you will come across many rather descriptive names, perhaps giving you the idea that these transistors must be radically different in many respects. As you can see from our description of the diffused transistor, the surface-barrier, and the micro-alloy transistor, one of the biggest differences is in the way the particular transistor is manufactured. The micro-alloy transistor, as an example, is sometimes referred to as a MADT—a

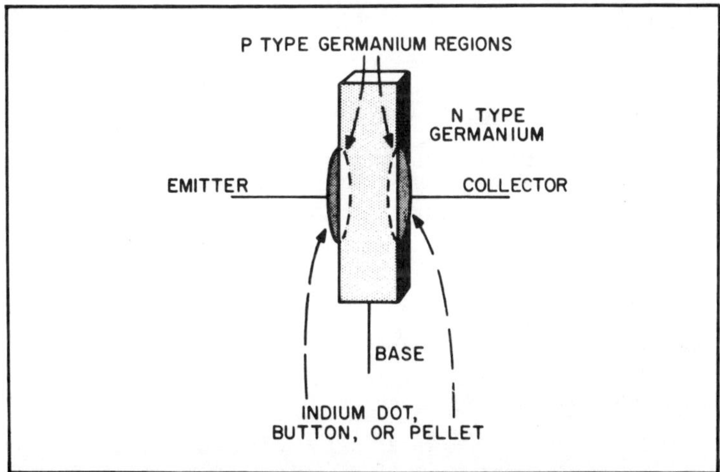

Fig. 8-5. The diffused transistor has an emitter and collector made of an acceptor element, indium. The indium, melted into the n-type material creates p-type material in the immediate region of the electrodes. This results in a pnp transistor.

micro-alloy diffused-base transistor. And, to compound the growing confusion, this same transistor is often called just an alloy-diffused transistor, or an alloy transistor. The alloy material, indium, is termed a dot or button, because of its shape on the surface of the n-type material base.

Manufacturers of transistors often use names of their own to describe transistors, sometimes creating the impression that there must be many different types of transistors. Thus, what we have called a diffused transistor is also called a diffused junction transistor, a fused transistor, a fused junction transistor, a diffused emitter-collector transistor, a diffused E-C transistor, graded diffused, or (quite strangely) drift transistors. The tetrode transistor, described earlier, is also known as a double-base junction transistor. These are not all, because there is no doubt that we must have missed a few, or somewhere there must be a manufacturer concocting still another name. If you come across a transistor whose name seems unusual (and you will) stop and reflect for a moment. It may be just the name that is new—not the transistor type.

JUNCTION TRANSISTORS

These are the workhorses in transistor radios and are the units we used in describing our audio, i-f and rf amplifiers. And, as we have emphasized so often, they come in two basic types—npn and pnp. A receiver may use npns, pnps, or both, in a single set.

Before you rejoice over the fact that the name for this very widely-used transistor type is so simple and easy to remember, it is our sad duty to inform you that like almost all other transistors it is available under a variety of names, the choice depending upon the semantic ability of the manufacturer's engineering department. Thus you may find the junction unit listed as a fusion alloy junction transistor or alloyed junction transistor. Either germanium or silicon can be used as the semimetal, and so this information is often included in the name. We will then get names such as silicon fusion alloy junction transistor or germanium fusion alloy junction transistor.

INTRINSIC TRANSISTORS

These are also known as intrinsic-region transistors, pnip transistors, npin transistors. Unlike the surface-barrier and diffused transistors in which we try to make the base as thin as possible, the intrinsic transistor may actually increase the base thickness and yet have an extremely high alpha cutoff frequency.

This may seem to contradict what we said earlier so let us examine Fig. 8-6 to see if we can come to some sort of reconciliation.

As you can see, in Fig. 8-6A, we have what appears to be an ordinary pnp transistor, but sandwiched in between the base and the collector, we have a block of material marked with the letter I. This letter I is the first letter of the word, intrinsic. The intrinsic region consists of a semiconductor that is free of any doping impurity. This means that the semiconductor has an equal number of holes and electrons, hence is neither n-type or p-type. Since its facilities for current flow are very poor, it has a high resistance.

The virtue of the intrinsic region is that it separates the collector and the base. This is like moving the plates of the capacitor further apart. When we do this, we reduce the capacitance. Earlier we mentioned that one of the hurdles a transistor must overcome is capacitance between electrodes. In an ordinary junction transistor, the base (input) and the collector (output) are an intimate contact. This makes the capacitance large enough to lower the alpha cutoff frequency considerably.

There is probably one statement that is now worrying you and that is the high resistance of the intrinsic region. If it is, think of the resistance that appears between cathode and plate of the ordinary vacuum tube. In power tubes or power transistors, where we work with high values of current, resistance between elements is important and we want it to be as low as possible. But for rf work, where the current in a transistor (or in a tube, for that matter) is so very small, high resistance does not seem to be detrimental.

The intrinsic transistor comes in two types, as shown in Fig. 8-6A and 8-6B.

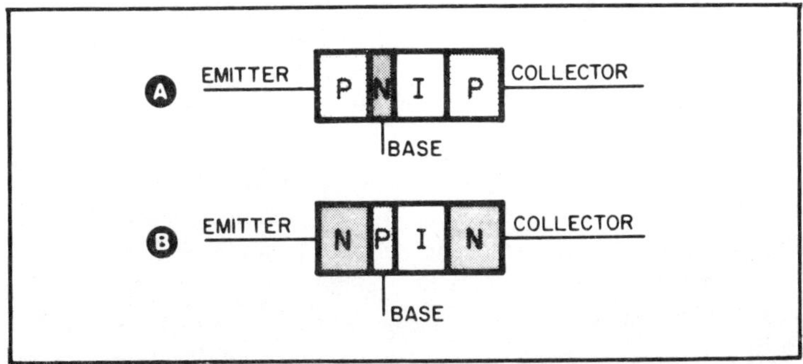

Fig. 8-6. An intrinsic transistor raises the alpha cutoff frequency by reducing capacitance between collector and base.

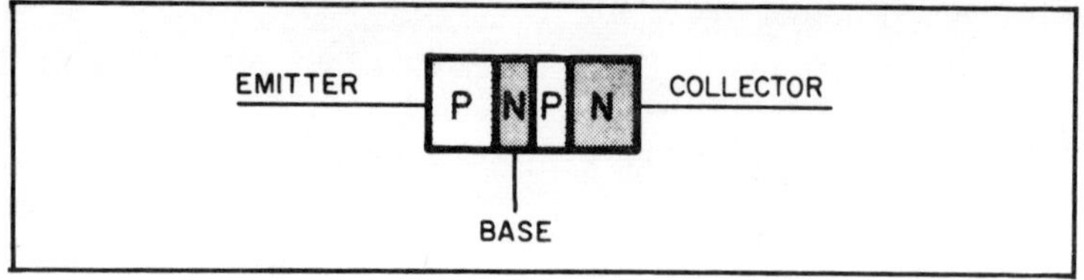

Fig. 8-7. Hook transistor resembles combined pnp and npn transistors.

HOOK TRANSISTORS

Every cook has a specialty, and those that feature soup, come up with some surprising (and often tasty) combinations. The true culinary artist will put anything and everything into the soup (at least once) just to see what will happen. So too in electronics do we have engineers who will try anything once. The more successful of these artists graduate to transistor design, where we can recognize their deft hands in the manufacture of unusual transistor-combinations. One such is known as the hook transistor, or four-layer triode.

We have an example of this transistor in Fig. 8-7. At first glance it looks as though we have a pair of pn diodes, but we can wipe our glasses and look again. Now you can see that what we really have (reading from left to right) is a pnp transistor or (starting at the base) an npn transistor. If this shocks you, consider a cathode circuit in which we go from the cathode to the plate of one tube, and then immediately into the cathode of the next.

Getting back to the hook transistor, the chief difference is in the way the collector is constructed. It has a slice of p-type material sandwiched in between itself and the base. With this sort of an arrangement, the hook transistor is most unusual in that it has an alpha greater than 1. No other transistor has such a characteristic except the point contact, the original transistor to be developed and which is now obsolete. Unlike the usual grounded emitter transistor arrangement, the hook transistor has an output which is in phase with its input.

Now let's consider the npn side of this unusual combination. Starting at the extreme right we have n-type material and we have marked this as collector for the entire hook transistor, but right now let's think of it as the emitter for the npn. From here we pass to a base which has no connection, and then to the n-material marked base but which is really the collector for our npn transistor. Since this is all very confusing, suppose we separate the npn section as in Fig. 8-8. In a typical npn transistor we would have a negative

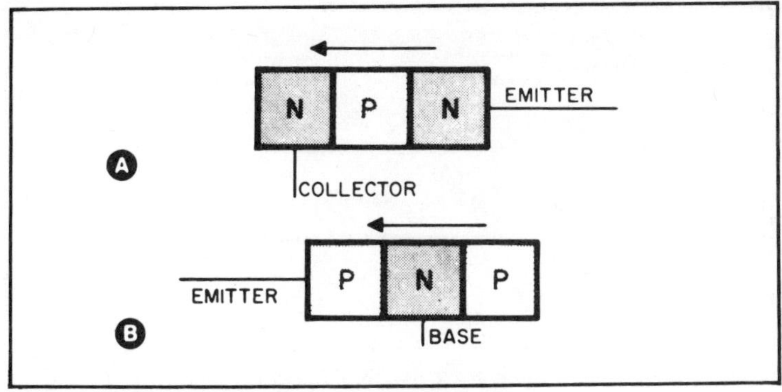

Fig. 8-8. The hook transistor can be separated into an npn, as in A, and a pnp, as in B.

voltage at the emitter and a positive voltage at the collector. But this is exactly what we have, since the collector in Fig. 8-8 is the base of the hook transistor and so really has a positive voltage on it. We haven't put any voltage on the p-section. This acts as the base of our npn unit, but it isn't biased in the usual manner. Because of the thinness of the base, electrons have no problem in flowing from the emitter to the collector. Note the direction of current movement for the npn transistor. As you know, in an npn transistor, current moves from emitter toward collector, and so the drawing in Fig. 8-8A fulfills this requirement.

In Fig. 8-8B, we have the remainder of the hook transistor. This is the pnp section and, as usual in this type of transistor, current flows toward the emitter. Now if we recombined our two transistors into the single unit of Fig. 8-7, we have two currents, and the output coming out of the emitter is the sum of the two.

THE POINT-CONTACT TRANSISTOR

Now that we have covered a few (and we mean just a few) of the possible transistor types, it is our sad duty to tell you that we have thus far completely ignored and neglected the honorable ancestor of all transistors—the point contact. Its glory dimmed by its illustrious offspring, the point contact may yet come back into its own since it has some remarkable characteristics. It has an alpha greater than 1, and it is adept where a transistor is needed for rapid switching purposes.

We have a drawing of the point-contact transistor in Fig. 8-9 and a photo in Fig. 8-10. As you can see, we've started with a slice of n-type material and have placed two springy metallic points on it.

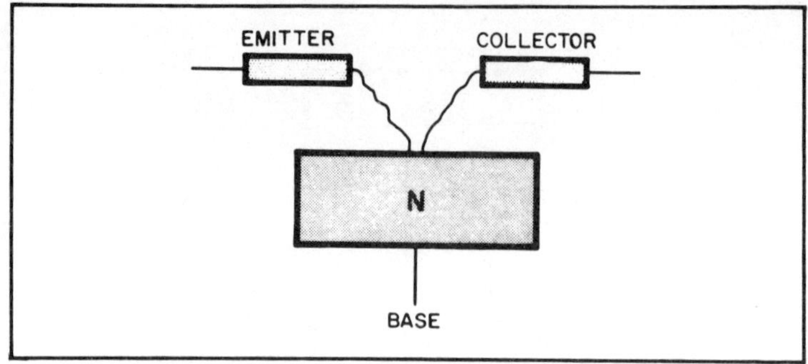

Fig. 8-9. Point-contact transistor is the forerunner of the junction type.

One of these is called the emitter, the other the collector, while the n-type section is the base. A voltage is placed across the emitter and collector leads momentarily, so that a current passes between them. This is believed to form p-type material regions in the immediate area around the contacts, thus supplying us with a pnp-type transistor. In use, the point contact is biased in the same way as a pnp junction-type transistor.

It is quite some time since we talked about alpha. Briefly, it is the ratio of collector current to emitter current and for transistors (except the hook transistor and the point contact) is less than 1. In other words, collector current is always less than emitter current. In the point contact transistor, collector current is sometimes as much as twice the emitter current. This might seem a great advantage, but it leads us into a trap. Current flowing through a transistor

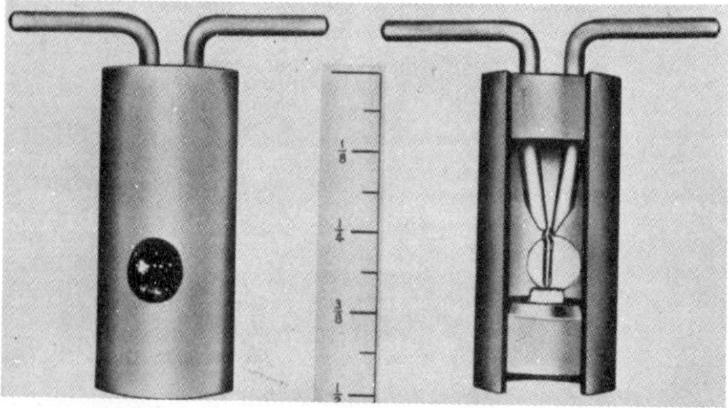

Fig. 8-10. Exterior and interior views of a point-contact transistor. (Bell Telephone Laboratories, Inc.)

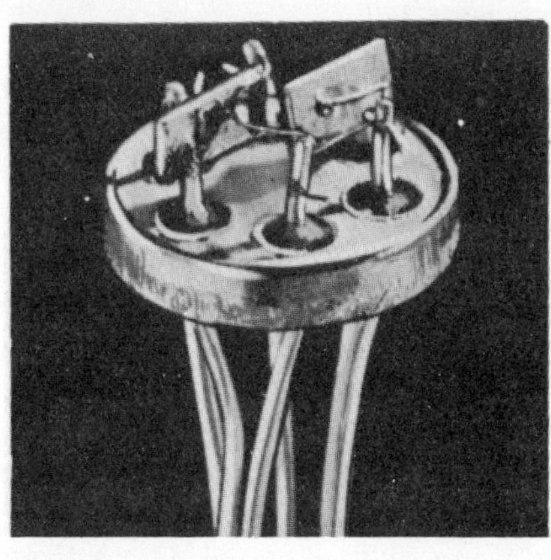

Fig. 8-11. Interior view of a multi-headed transistor. (Electronic Transistors Corp.)

(uncontrolled) tends to produce more current. This is current runaway and can destroy a transistor faster than you can reach for a switch to turn off the current.

MULTI-HEADED TRANSISTORS

Ever pack a suitcase? With a little ingenuity, planning and some brute force to close the cover, the amount of clothing you can get into a small space is amazing. A similar technique is used in vacuum tubes, ranging all the way from duo-diodes, diode triodes,

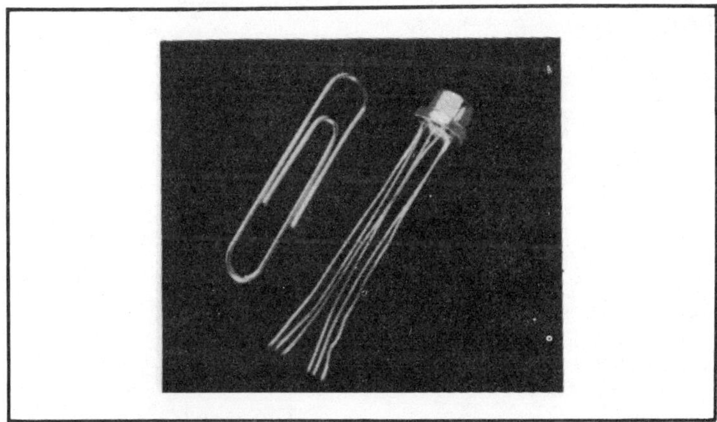

Fig. 8-12. Multi-headed transistor, shown in actual size. This unit contains two transistors in one package. (Electronic Transistors Corp.)

triode-triodes, right on up to the Compactron—a multiplicity of tubes all within the single glass envelope.

Transistors aren't to be outdone by this. Multi-headed transistors can be had in any desired combination. We have an inside view of one of these in Fig. 8-11. This unit contains two transistors in one package. The two transistors are completely independent of each other, can be used singly or in combination. Each multi-headed unit has six leads. You can get an idea of the size of the unit in Fig. 8-12.

TANDEM TRANSISTORS

In some vacuum tubes certain elements are internally connected; in others, each element is independent. Thus, we can have a suppressor grid internally connected to the cathode, or wired to a pin in the base so we can make any use we want of the suppressor.

The multi-headed transistor is designed for external wiring, but as shown in Fig. 8-13 we can also have transistors connected internally. The unit in Fig. 8-13 has the emitter lead of the first transistor wired to the base of the following one. In some tandem units, the collector is the tied-in element, wired to the base of the following transistor. All sorts of combinations are possible but aside from the saving in space, there is no particular advantage in using a tandem transistor instead of two separate units.

GERMANIUM AND SILICON

In an earlier chapter we brought in a little bit of chemistry for you to examine. At that time we became particularly interested in those elements which had four electrons in their outermost orbit. We have a nice collection of these—carbon, silicon, germanium, tin, and lead. Now the order in which we have listed these elements isn't accidental. Carbon has two electron rings, silicon has three, germanium has four, tin has five, and lead has six. Thus we started with the lightest element and moved along to the heaviest. Although

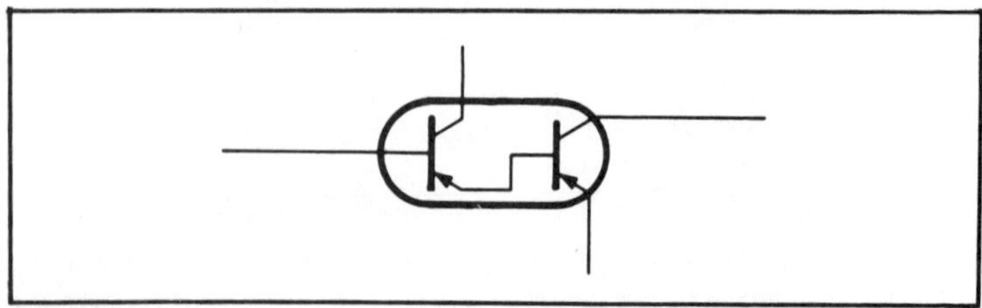

Fig. 8-13. The tandem transistor is connected internally.

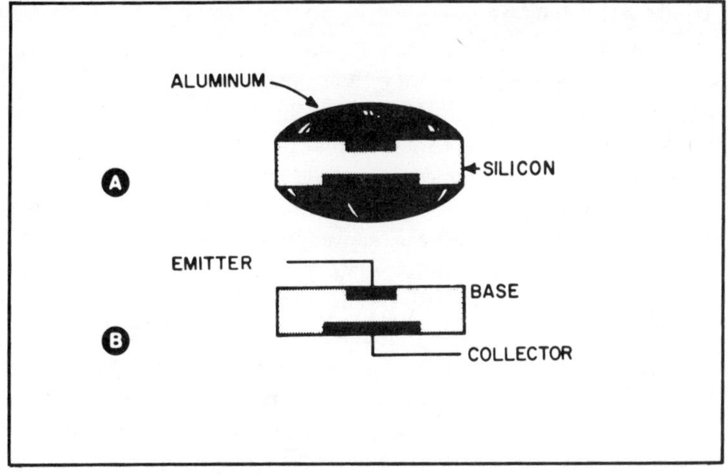

Fig. 8-14. The evaporation-fused transistor uses aluminum, fused into the silicon, for the emitter and collector connections.

all of these elements are related in that they have similar outermost electron rings, they do have very definite differences. Consider the two heaviest—tin and lead. These are metals, nor do you have any doubt that they are. Silicon and germanium, though, aren't really metals, but semimetals. And when we get to carbon we arrive at an element that does not have metallic properties.

Germanium was originally the most widely used semimetal or semiconductor. Silicon transistors, though, do have certain characteristics that make them more desirable. They do not suffer as much from high temperatures as germanium transistors. Also, they have lower leakage currents. Silicon is a much more abundant semimetal than germanium, but it is much more difficult to purify.[1]

As in the case of germanium, silicon must be purified and then "doped" with controlled impurities so that we get either p-type or n-type silicon. One type of silicon transistor, known as the silicon alloy or evaporation-fused transistor, is shown in Fig. 8-14.

In the manufacture of the silicon-alloy transistor, cavities are cut into the silicon wafer. Aluminum is evaporated and allowed to alloy into the surface of the silicon. The result of this process is shown in Fig. 8-14A. The excess aluminum is then removed. The aluminum that remains in the upper and lower cavities forms the emitter and the collector, illustrated in Fig. 8-14B.

---

[1] The nonmetal carbon, and the semimetal silicon, as elements or as compounds form all living material. Practically all of the earth's minerals are accounted for by just these two elements.

TRANSISTOR DIMENSIONS

Transistors, other than those used for work as power amplifiers, are well-known for their smallness. Typical units are shown in the photograph, Fig. 8-15. But small as they are, even tinier units are needed for special applications where all space is at a very high premium. This could include missile units or computers using tremendous numbers of transistors, or in hearing aids. Figure 8-16 shows some of these, representing the notes on a musical staff. Actually, these subminiature transistors, intended for hearing-aid use, are smaller than the musical notes they will amplify.

At this point we must be careful not to fall into the trap of thinking that reduction in size means an increase in the alpha cutoff frequency. With vacuum tubes, one way of raising the operating frequency was to bring the cathode closer to the plate, and since the tube envelope accommodated this space shrinkage, the result was a smaller tube. This situation is not analogous in the case of transistors. In some cases, subminiature transistors have *lower* alpha cutoff frequency than their bigger brothers.

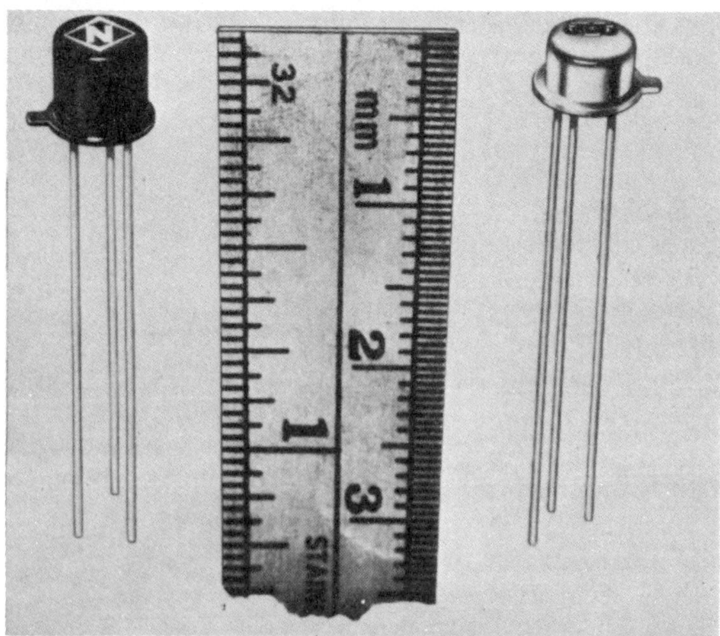

Fig. 8-15. These silicon mesa transistors are about 3/16 inch long (National Semiconductor Corp.)

Fig. 8-16. The transistors shown here in actual size are smaller than the musical notes they represent. (Raytheon Co.)

A representative subminiature might have a height of 0.160 inch compared to 0.260 inch for the more common type. This is a difference of only a tenth of an inch, but think of it in terms of percentage and you will see that the decrease in size is quite astonishing. But this doesn't mean that the end has been reached. The hearing-aid transistors in Fig. 8-16 are less than 0.1 inch in height, while some are down to 0.05.

TRANSISTOR MANUFACTURE

The first step, and probably the most important in the manufacture of transistors of all types, is refining the semiconductor material. The refining process is carried to the point where the semiconductor is almost 100 percent pure. The silicon or germanium is then melted. While the semiconductor is in this molten state, either donor or acceptor impurities are added. The type of impurity will determine whether we will have n-type or p-type semiconductor.

At this time, a small bit of crystalline material known as the seed is dipped into the melt and is then withdrawn slowly. The effect of the seed is to trigger the crystalline formation of the semiconductor. This crystalline formation is not in the direction of millions of tiny crystals, but rather more like that of a single large crystal.

Did you ever watch a salami being sliced by a machine? That's our next step, but the cutting of the semiconductor crystal is carried on in sanitary surroundings that no delicatessen could duplicate. Each thin semiconductor wafer is then cut into tiny chips. These tiny bits of semiconductor crystal are our transistors. In the manufacture of the most common type of transistor, the junction transistor, tiny dots of indium, aluminum or antiomony are alloyed to the surfaces of the semiconductor.

Our transistor is now completed, but we must somehow make connections to the emitter, collector, and base. As a start, we mount the transistor on a metal tab, as shown in Fig. 8-17. This tab now forms our connection to the base of the transistor. As you can see in the drawing, a hole is cut in the tab to permit access to the dots of indium. The unit is now mounted in its housing or package and is ready for testing.

In Fig. 8-18 we have a cutaway photo of an alloyed-junction transistor. As you can see, this resembles the drawing in Fig. 8-19. However, various manufacturers all have their own techniques. Thus, a similar transistor in Fig. 8-20 shows wires connected to the indium dots. These wires help support the transistor and also make

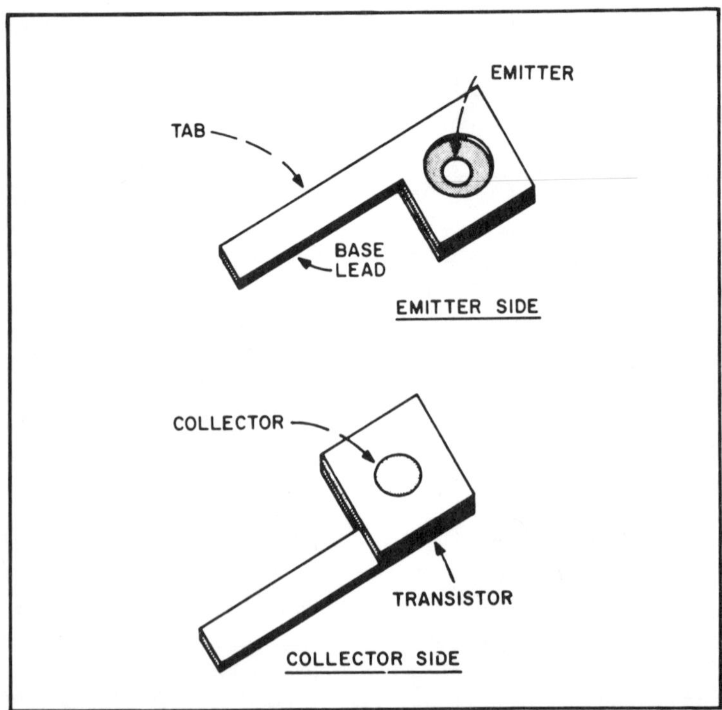

Fig. 8-17. The transistor is mounted on a tab which forms the lead for the base.

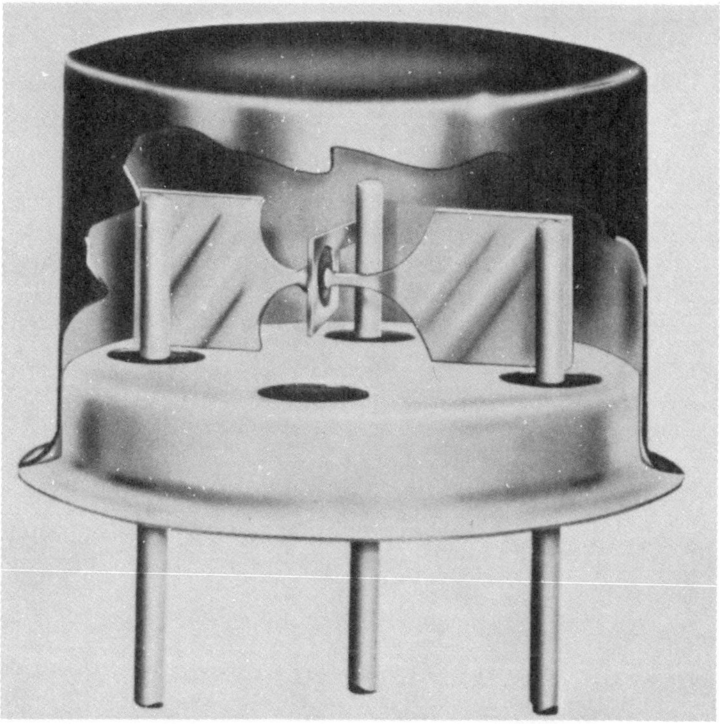

Fig. 8-18. Interior view of an alloyed-junction transistor. A pair of "bow ties" connects the emitter and collector to their external leads. The center, or base lead, is connected to the body of the transistor by a tab. (Raytheon Co.)

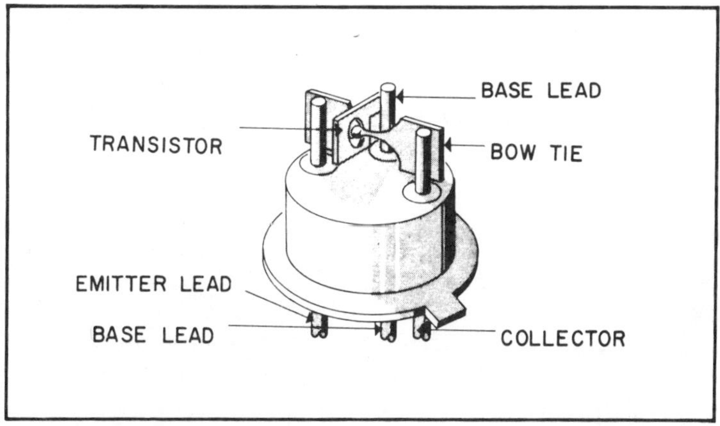

Fig. 8-19. The bow ties connect to the emitter and collector. The tab (see Fig. 8-17) connects to the base lead.

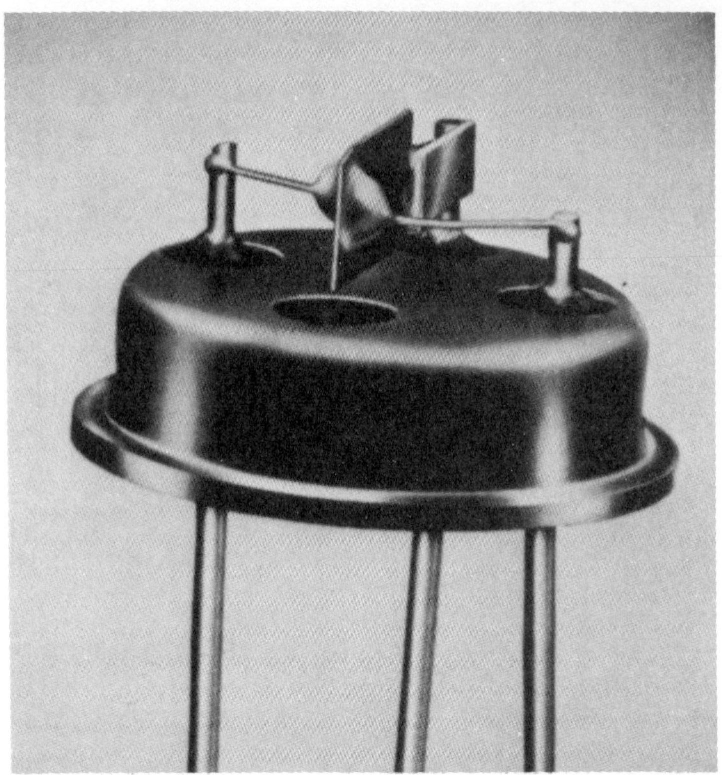

Fig. 8-20. In this alloyed-junction transistor, a pair of wires is used to help support the transistor and to connect to the emitter and collector. Note the circular holes cut in the "package" to permit the external leads to come through to the emitter, base and collector. (Raytheon Co.)

connections to the external emitter and collector leads. No tab is used in this particular construction with the exception of a small rectangular block from the body of the semiconductor to the base lead.

POWER TRANSISTORS

Power transistors, as their name implies, represent the final link between the load and the signal. In the case of a radio receiver the load is the speaker and since the voice-coil type of speaker is a current-operated device, it must be driven by a transistor capable of delivering audio power. Power transistors are no different than ordinary transistors, except they are specially designed to handle large values of current. Thus, in the case of ordinary amplifying transistors we are accustomed to currents in microamperes and milliamperes; for power units we can have dc collector currents of 10, 20 amperes or more. These large currents raise immediate problems, for where we have a high current there is always the pressing question of an IR drop. When currents are measured in amperes, just a small amount of resistance causes a voltage loss. Of equal concern is the fact that, large amounts of heat result from I^2R losses. Since the power to be dissipated (in the form of heat) varies as the square of the current, heatsinks must invariably accompany power transistors. Thus, if we consider the movement of a moderate amount of current such as 10 amperes through a 1-ohm resistor, we would have (based on the formula: $W = I^2R$) 10 × 10 × 1 or 100 watts. Now this may not sound like much, but try touching a 100-watt electric-light bulb after it has been on for a few minutes and you will get the skin-blistering idea.

NUMBER OF LEADS

In Fig. 8-21 on the next page we have the step-by-step processes used in the manufacture of power transistors. The heatsink is an integral part of the transistor assembly, but in itself this is often inadequate. The metal chassis on which the transistor is mounted helps dissipate the heat, or specially-designed separate heatsinks are provided. These are often corrugated or ribbed to supply the maximum amount of radiating surface.

Not only are power transistors much heavier and larger than ordinary amplifying transistors, but instead of having three connecting leads most have just two. These leads are for the base and the emitter. The collector of the power transistor is mechanically and electrically connected to the metal shell of the unit. Incidentally, as you might expect, while the majority of amplifying transistors (other than power transistors) have three leads, there are some that have four. A transistor of this type is the post-alloy diffusion transistor (abbreviated as PADT) which has an extra lead connected to the case.

MOUNTING TRANSISTORS

Since transistors (other than power types) are so extremely small and light, mounting them is no more of a problem than mounting a quarter-watt resistor. Just keep in mind that transistors aren't too partial to the heat of a soldering iron and there should be some sort of thermal shunt on the transistor lead, between the tip of the soldering iron and the transistor proper, to bypass the heat of the iron. You can avoid the need for a thermal shunt by using a clean,

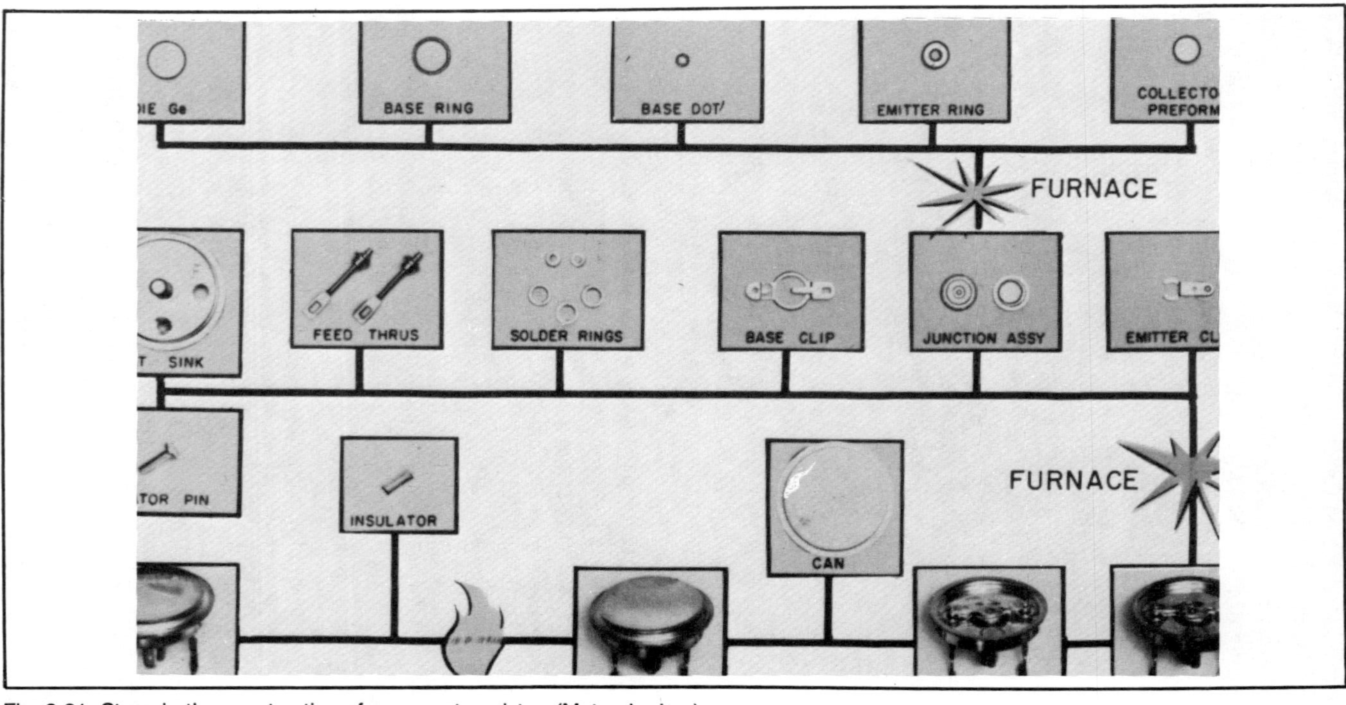

Fig. 8-21. Steps in the construction of a power transistor. (Motorola, Inc.)

well-tinned iron and soldering with the correct melting point solder. Transistors are not affected by the rather substantial magnetic field surrounding a soldering gun or the smaller field around the usual soldering iron.

SOCKETS

Sockets are sometimes used for transistors, but since the life

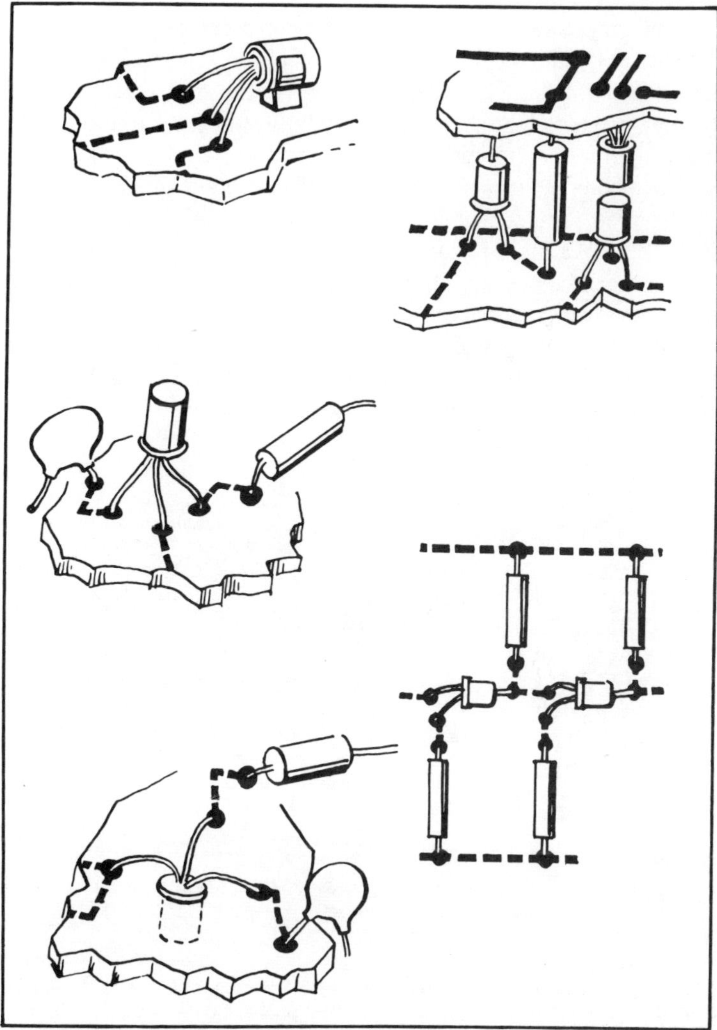

Fig. 8-22. These are some transistor mounting techniques. Transistors can be placed in any position. (Raytheon Co.)

expectancy of a transistor is equal to or better than most other components, the transistors can be soldered directly into place. Unlike tubes, transistors can be soldered in any position. They can be mounted vertically, horizontally, below, above or between printed boards, as shown in Fig. 8-22. Transistors can be connected so that they sit on their leads, or hang from them.

CIRCUIT PACKAGING

When you service a radio or television receiver you do so with the idea of finding the one component that is defective. Granted that sometimes the trouble is caused by more than one defective part, the fact remains that servicing consists first in localizing the difficulty and then in isolating the troublemaker.

For equipment used in industry and in military applications this approach is time-consuming. It can also be expensive, since the down time of a machine such as a computer can cost far more than any amount of servicing. For this reason a circuit that does not work properly is completely replaced (Fig. 8-23). Granted that a half-dozen or so good parts might surround one weak component, this is still faster and far more economical than to institute a search for and replacement of an individual resistor, capacitor, or transistor.

Fig. 8-23. Plug-in modules often contain complete stages. It is a form of construction that supplies a very high component packaging density. In the instrument shown above, components are mounted on boards which slide into grooves. Connections are made through special connectors on the inside edge of each of the boards. (EI Labs)

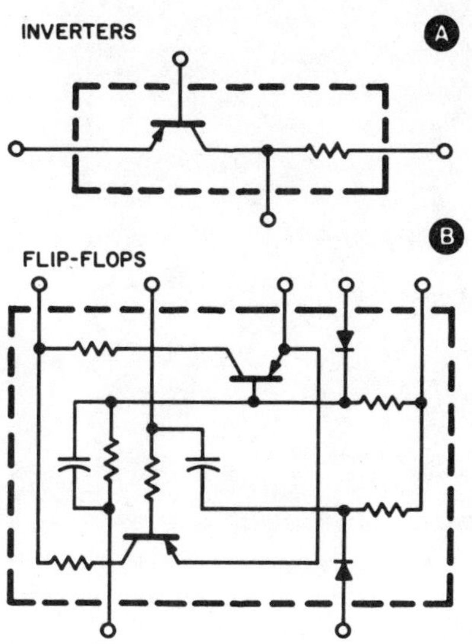

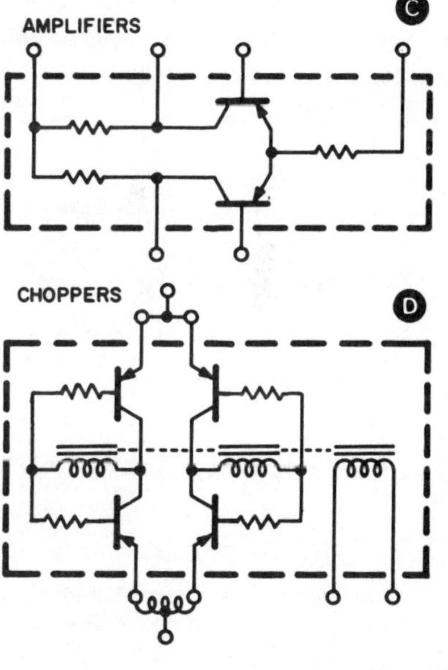

Fig. 8-24. Modules are taking the place of much electronic wiring. Some modules are no bigger than a transistor alone. Computers and data control equipment may consist of a dozen module types repeated several hundred times. One group of modules is logic circuits which have multiple inputs. A signal or the lack of a signal at one or more of these inputs will cause or prevent an output signal to the next stage which may be another logic module, an inverter, a flip-flop of one of many varieties or a relay or readout device.

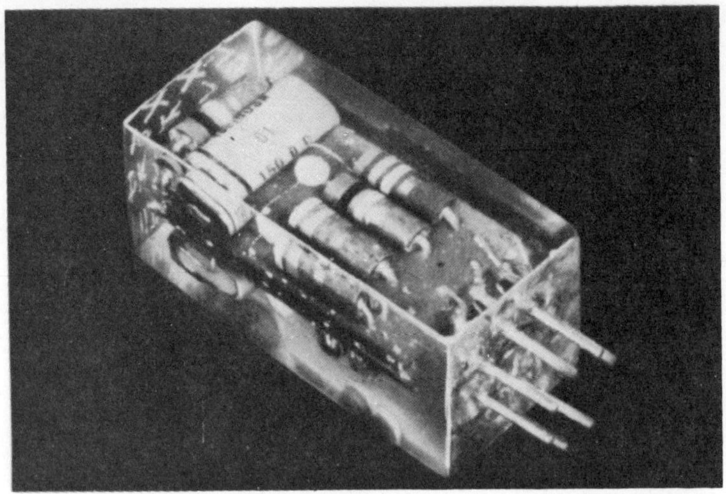

Fig. 8-25. Packaged transistor circuit. (Bell Telephone Laboratories, Inc.)

Plug-in modules are simply switched in the field, and defective units sent back to a central repair facility where the bad component is found. Then the "card" is good again—ready to replace some *other* bad one!

In Fig. 8-24 we have a number of packaged circuits. These can be used in original design applications or as replacements. A complete electronic device can be built of these blocks and can be repaired by simple substitution. The components are connected and are then encapsulated and potted into a variety of packages—tubular, rectangular, bathtub shape, etc. Connecting leads are brought out to a 9-pin standard plug-in base, octal and subminiature in-line base, a 7-pin standard base, or with pig-tail leads or pin terminals. Replacement is as simple and as easy as replacing a tube. Figure 8-25 shows an experimental model of a transistor amplifier for telephone switching circuits. All components of this miniature plug-in type device, including the beaded-junction transistors, are compactly embedded in lucite. The model is about 1⅝-inch long. Some such solid-state devices are available that directly replace tubes!

OTHER TRANSISTOR TYPES

The transistors we told you about earlier in this chapter are just a few of the many different types. There are so many of them now, and so many constantly being developed, that a book could be written just describing them. And each new type that is produced may have as many as a half-dozen different names depending on the

Fig. 8-26. Integrated circuits come in a variety of different packages, but they all share one feature: extreme compactness.

manufacturing techniques being used. Field-effect transistors are another matter. We will look closely at these in Chapter 9.

There is one more item we should discuss and that is that transistors represent just one branch on the semiconductor tree. There are many types of semiconductor devices. Among the best known of these are the diodes and integrated circuits.

Transistors are probably best known to you through their use in radio receivers and transmitters, since we have examined these circuits now in some detail. But transistors are the heart of many, many kinds of greatly miniaturized circuit packages known as integrated circuits (ICs).

A circuit that (in the tube era) would have occupied an entire house—including the baths!—now may be built with a few "chips" no larger than pencil erasers. A circuit that would have required a specially installed power line just for the purpose of heating its tube

Fig. 8-27. Integrated circuits lend themselves nicely to printed-circuit construction. This layout would have been considered fantasy just a few years ago. But electronics is, itself, a very young science.

filaments, in the early days of electronics, can now be powered from a few penlight cells! Integrated circuits (Figs. 8-26 and 8-27) are truly the eighth wonder of the world. Even a small computer, called a microcomputer or microprocessor, can be encapsulated within one single, tiny IC.

Various kinds of digital logic circuits have been developed for a wide range of applications. Integrated circuits are well suited for this kind of thing. These digital ICs may contain hundreds of individual diodes and transistors. The subject of digital logic could well occupy several volumes, so we won't dig into that here. But the whole business would never have been possible if not for that modest, quiet, efficient little transistor, that guy who started it all.

QUESTIONS

1. What is transit time? How does it affect transistor operation?

2. What is alpha cutoff frequency?

3. How does the tetrode transistor differ from the triode transistor?

4. How does the surface-barrier transistor overcome the problem of transit time?

5. What are the differences between a micro alloy transistor and a MADT type?

6. Describe the main advantages of silicon over germanium as a semiconductor.

7. What does the hook transistor have in common with the point-contact transistor?

8. What is an integrated circuit? What advantages do ICs offer over individual components?

9. What is a modular piece of electronic equipment? How does the modular type of assembly reduce down time?

Chapter 9

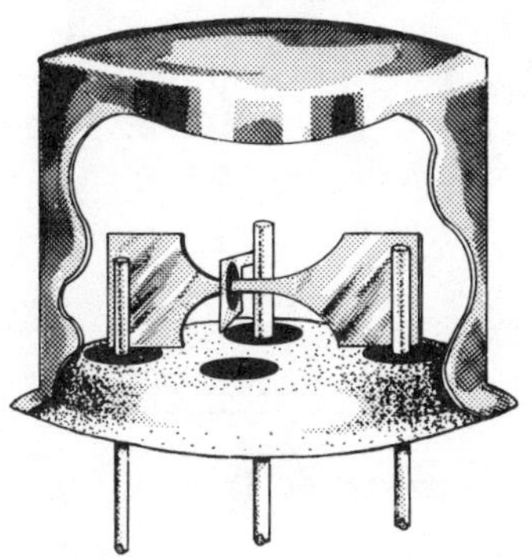

The Field-Effect Transistor

WE HAVE NOW SEEN SEVERAL DIFFERENT TYPES OF TRANSISTORS. Of these, the npn and pnp are the most common. They are called bipolar transistors because they contain two junctions: the emitter-base and collector-base pn junctions.

In recent years, however, a new type of transistor, called the unipolar or field-effect transistor (FET), has come into increasingly common use. There are many different kinds of FET. They exhibit characteristics that make them better for certain applications than bipolar devices. One such advantage is that the FET is generally less noisy. The constant motion of electrons in any circuit creates electrical noise; the motion of the molecules caused by heat causes thermal noise. Sometimes it is necessary to keep this noise down as much as possible. For example, we may want to build a very sensitive receiver for vhf or uhf; any electrical or thermal noise in the first stage (front end) will limit the ultimate sensitivity of the unit. Field-effect transistors are great guys when it comes to noise. They do their jobs with hardly a whisper.

FETs are also used in some linear-amplifier circuits. They can be used as oscillators, too. Let's begin our discussion by examining the structure of a simple FET, called the junction field-effect transistor (JFET).

Field-effect transistors are all made from the same p-type and n-type semiconductors as are bipolar devices.

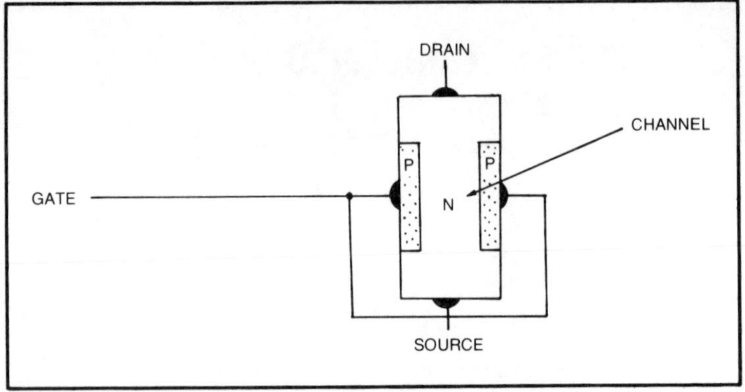

Fig. 9-1. Cross-section of an n-channel JFET. The current flows from the source to the drain through the n-type material. The two sections of p-type material form the gate.

CONSTRUCTION OF A JFET

In Fig. 9-1, we see how an n-channel JFET is put together. This particular FET has a region of n-type silicon sandwiched in between two sections of p-type material, as shown. Electrical leads are connected to each end of the section of n-type silicon, called the channel. Leads are also connected to the two p-type sections, which are called the gates. The gates are usually tied together inside the housing of the FET. (However, they may be left separate for certain special-purpose devices.) Figure 9-2A shows the schematic symbol for an n-channel JFET. The electrodes at either end of the channel are called the source and drain; they are interchangeable. The gate electrodes are called—what else?—the gate! The JFET need not have n-type material as the channel, although we will restrict our discussion of JFET operation to n-channel types. The gates may be made of n-type silicon and the channel of p-type silicon; in this case,

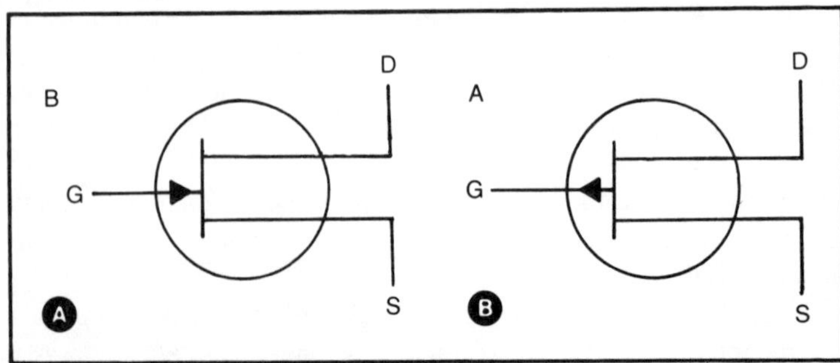

Fig. 9-2. Schematic symbols for the JFET. At A, n-channel; at B, p-channel.

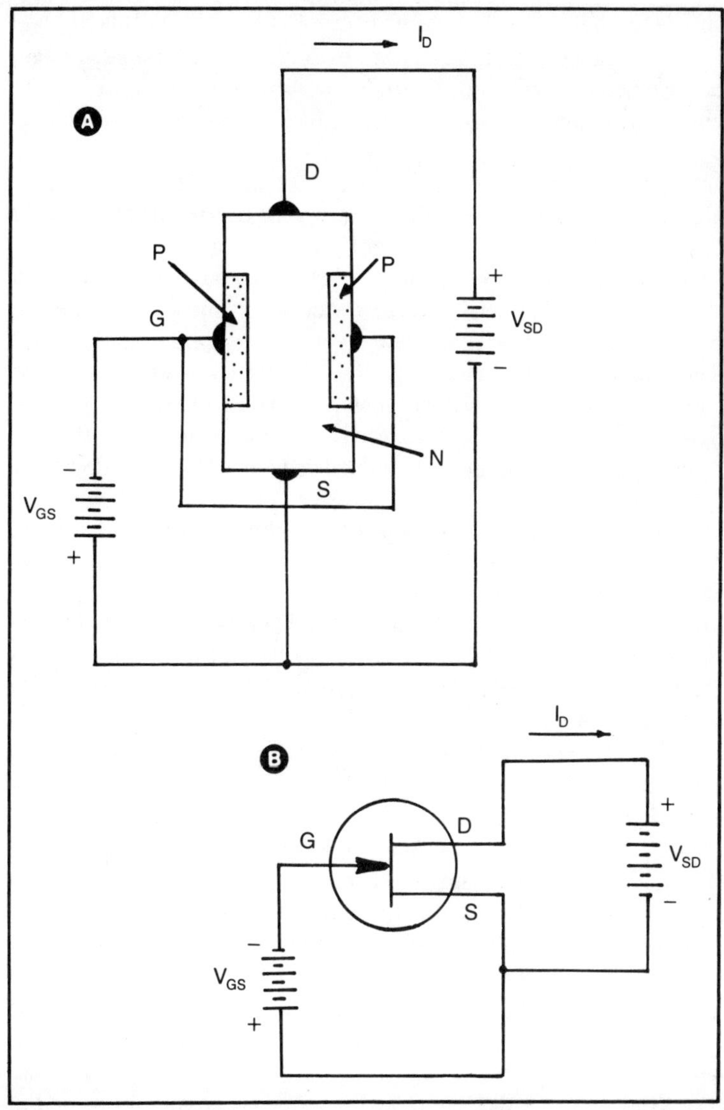

Fig. 9-3. Biasing arrangement for the n-channel JFET. The gate is negative with respect to the source. The drain is positive with respect to the source; this is shown at A in pictorial form and at B in schematic form.

we have a p-channel JFET. Its schematic symbol is shown in Fig. 9-2B.

Let us now consider the characteristics of the n-channel device. Imagine that there is a small battery, of variable voltage, connected between the gate and the source. We call this voltage V_{GS}. Now suppose there is another battery, also of variable voltage, between the source and the drain; we call this voltage V_{SD}. The circuit arrangement, both pictorial and schematic, is shown in Fig. 9-3.

When $V_{GS} = 0$, that is, the gate is not connected to anything at all, a certain current I_D will flow through the JFET and appear in the drain circuit. (Actually, the polarity of V_{SD} may be reversed and things will remain just the same. There is no pn junction between the source and drain; however, we call the terminal with the more positive potential the drain by convention in an n-channel JFET.) The larger we make the voltage V_{SD}, the more current I_D will flow through the device. The relationship is linear as long as we do not make V_{SD} too great. Therefore, the JFET behaves like a resistor. Its resistance may be simply determined by Ohm's Law: Divide V_{SD} by I_D.

Now suppose we apply a positive voltage to the gate, so that $V_{GS} = +1$. This constitutes forward bias of the pn junction between the source and the gate. What will happen? Obviously, some current will flow from the source to the gate and out the gate lead. Thus I_D decreases. But the JFET is not designed for this kind of operation. If the gate circuit draws current, we cannot make the JFET work very well as an amplifier.

So let's see what happens when we put reverse bias at the gate-source junction. Suppose $V_{GS} = -1$. Figure 9-4A illustrates the resulting situation inside the JFET with small values of source-drain voltage. The reverse bias produces a depletion region in the n-type material comprising the channel, as we expect with any pn junction. This makes the effective channel narrower. Therefore, its cross-sectional area is reduced, and this increases its resistance. And, of course, more resistance means less current; thus I_D goes down. If we keep $V_{GS} = -1$ and adjust the voltage V_{SD}, we will find that I_D still varies in linear proportion to V_{SD}, as long as V_{SD} is not too great.

Now suppose that we put even more negative voltage on the gate, thereby reverse-biasing it further. Suppose $V_{GS} = -2$. This will cause the depletion region in the channel to grow wider, and the current I_D will get still smaller since the effective channel has become more narrow. If we keep increasing the gate voltage to greater and greater negative values, the depletion region will eventually get so wide that the channel is totally closed (Fig. 9-4B). Then, no matter what V_{SD} is, the current I_D will remain zero. We call

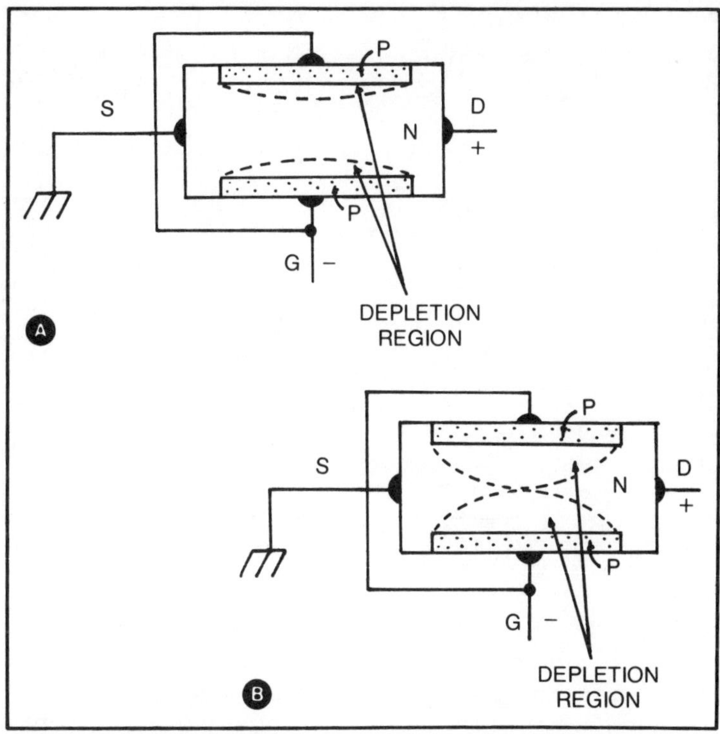

Fig. 9-4. As the gate is made more and more negative, the depletion, the depletion regions in the n-type channel become larger. At A, we see considerable narrowing of the channel. At B, the channel is pinched off. The source-drain potential here is rather small.

this situation pinchoff, and the voltage required to cause pinchoff is known as the pinchoff voltage.

A JFET thus acts, as you can see, like a variable resistor. The resistance is controlled by the voltage V_{GS}; for a given, constant reverse-bias voltage, the relation between I_D and V_{SD} is always linear as long as V_{SD} doesn't get larger than a certain value. The variable-resistance effect is shown by Fig. 9-5. In Fig. 9-6, we see the linear relationship between I_D and V_{SD} for various values of V_{GS}.

This may all seem a bit confusing, with these strange voltage notations. Just remember that S stands for source, G stands for gate, and D stands for drain in the subscripts.

Suppose we could probe inside the JFET with a voltmeter, connecting the negative terminal to the source and the positive terminal to various points in the channel. We would then see a zero voltage, of course, when the positive terminal was right next to the

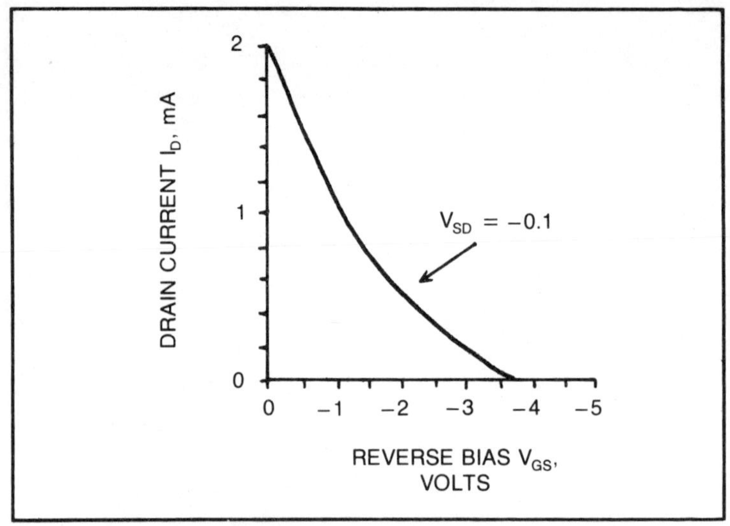

Fig. 9-5. Drain current as a function of gate voltage for small values of V_{SD} (source-drain voltage), in this case −0.1 volt.

source, and V_{SD} with the terminal next to the drain. At intermediate points, we would see intermediate voltages, just as we would if we were to perform the same experiment with a carbon-composition resistor. But notice from Fig. 9-1 that the gate p-type material runs

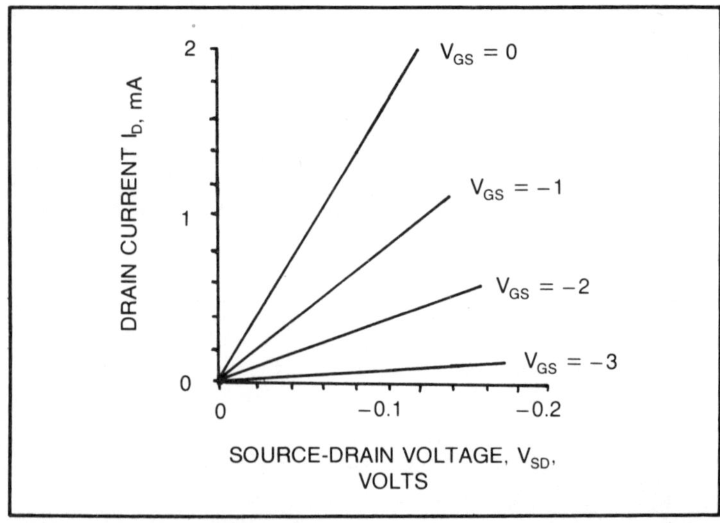

Fig. 9-6. Drain current I_D as a function of source-drain voltage V_{SD} for various values of gate voltage V_{GS}.

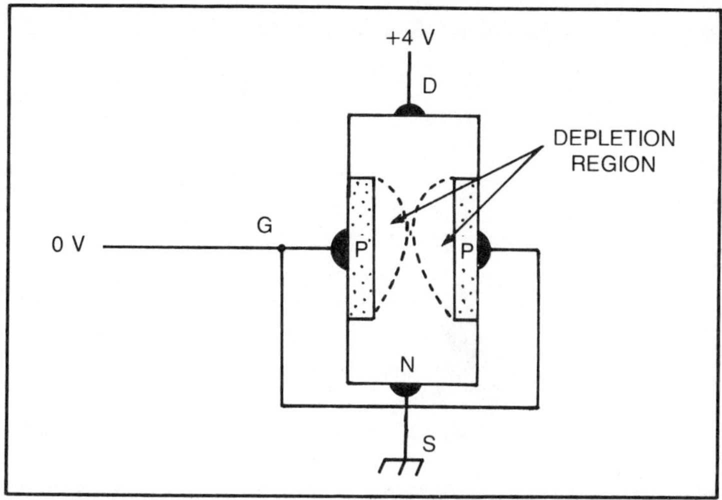

Fig. 9-7. Even when the gate bias is zero, large values of V_{SD} (in this case −4 volts) will cause pinchoff near the drain. This tends to limit the current that can be forced through the channel.

almost the whole length of the channel. The voltage between the gate material and the channel is more negative near the drain than near the source, even though V_{GS} is a constant value. This effect is not very important, in fact it is insignificant, when the voltage V_{SD} is small. But when V_{SD} is fairly large, the reverse bias is effectively much greater near the drain. Figure 9-7 shows the conditions inside the JFET when $V_{GS} = 0$ and V_{SD} is about −4 volts. While there is no reverse bias near the source, there is 4 volts reverse bias near the drain, and this is enough to pinch off the JFET channel. The current, I_D, is not completely cut off by this effect. For a given reverse bias at the gate, increasing voltages V_{SD} cause the current to increase and then level off as pinchoff is reached. Figure 9-8 illustrates the relation among I_D, V_{GS}, and V_{SD} for large as well as small voltages V_{SD}.

The operation of a p-channel JFET is exactly the same as we have seen here, except that all the polarities (and ps and ns) are interchanged. It's just this way with bipolar transistors, too—the pnp and npn operate essentially the same way, but with reversed polarity.

INSULATED-GATE FETS

There are other ways to build field-effect transistors. Insulated-gate FETs are becoming more and more common. The

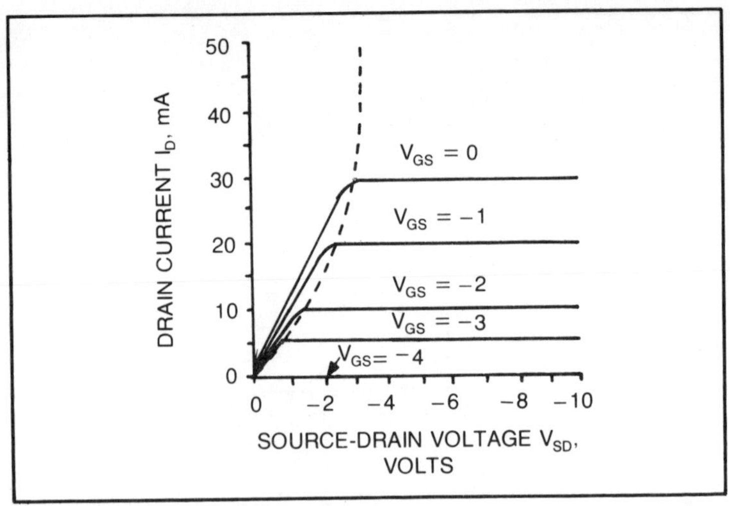

Fig. 9-8. Drain current I_D as a function of source-drain voltage V_{SD} for various bias voltages V_{GS} over a wide range of V_{SD}.

most often-seen of these is the metal-oxide silicon field-effect transistor. That's a mouthfull! So we shorten it to MOSFET. There are several different kinds of MOSFET; we may have n-channel or p-channel devices, and they may be operated in two ways, called depletion mode and enhancement mode. Furthermore, MOSFETs may have one or two gates. Let's look at the construction of an n-channel, one-gate MOSFET of the depletion-mode type first.

Figure 9-9 shows the way the MOSFET is built. This is a depletion-mode device; that is, it operates because of the narrowing of a channel by reverse bias, just as is the case with the JFET. (But you can see there are differences between this and the JFET!) The source and drain are connected to two pieces of n-type silicon, heavily doped. The channel consists of lightly doped n-type material, and it has a rather high resistance. All this n-type material is imbedded in a larger piece of p-type silicon. The base of the p-type block is connected to the source as shown, via a large metal (usually aluminum) electrode. The gate electrode is insulated from the channel by a layer of silicon dioxide.

Because of the insulating layer of silicon dioxide between the gate and the channel, the gate resistance is extremely high—trillions of ohms! Therefore the MOSFET gate circuit draws almost no current. This makes the MOSFET extremely valuable for high-impedance circuits.

When a negative voltage is applied to the gate electrode, a

depletion region forms near it, and this reduces the effective width of the channel. The result is the same as it is with the JFET; the drain current I_D is reduced because the cross-sectional area of the channel is smaller, resulting in a higher effective resistance. For small values of source-drain voltage, V_{SD}, the relation between I_D and V_{SD} is linear, just the way it is with the JFET. And also like the JFET, the MOSFET becomes pinched off if V_{SD} or V_{GS} get large enough. The characteristics are almost the same in every way, except for the higher input resistance of the MOSFET.

The MOSFET is ideal for weak-signal amplification because of the low input current drawn. A tiny bit of signal, arriving from many thousands (or perhaps someday millions) of miles, has no strength to provide current. Some MOSFETs have a gate on both sides of the p-type substrate, and these devices are called dual-gate MOSFETs. While the single-gate MOSFET behaves like a triode tube, the dual-gate MOSFET acts very much like a tetrode. Schematic symbols for n-channel and p-channel MOSFETs, both single gate and dual gate, are shown in Fig. 9-10.

The enhancement-mode MOSFET operates in a different way than the JFET or depletion-mode MOSFET. Rather than constricting the flow of electrons, the enhancement-mode device, as its name implies, acts to improve electron flow. Figure 9-11 shows the physical construction of an n-channel enhancement-mode MOSFET. Note that there is no channel built into the device. This

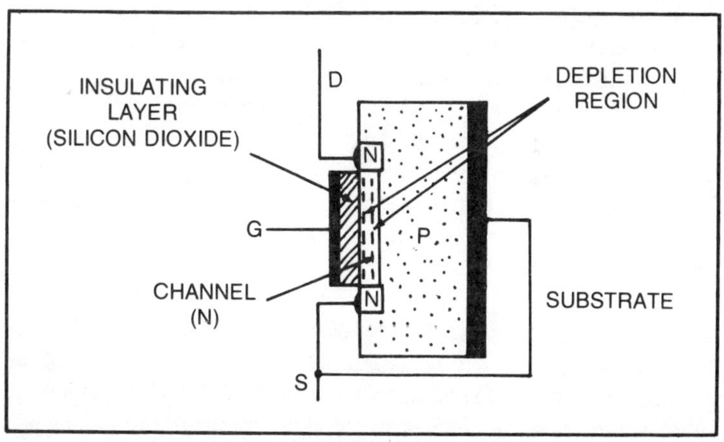

Fig. 9-9. Cross section of a MOSFET. This is a depletion-mode device. When a negative voltage is placed on the gate electrode, the n-type channel becomes narrower because of the depletion region. A p-channel depletion-mode MOSFET simply has the p-type and n-type materials reversed, as compared with this n-channel device.

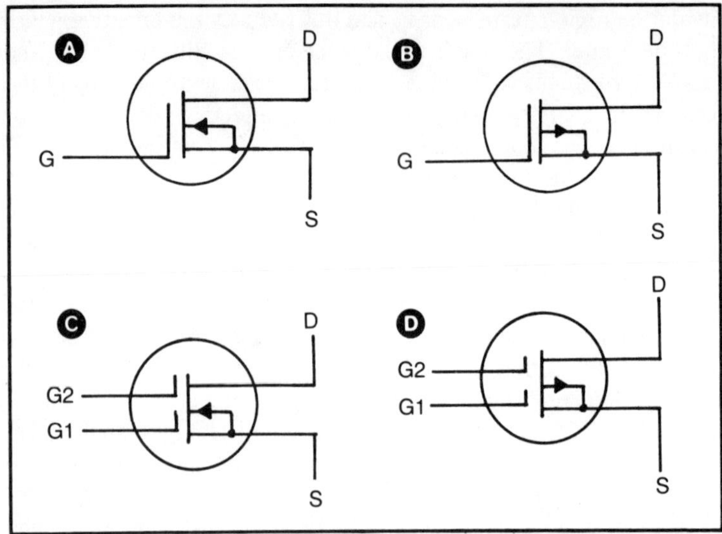

Fig. 9-10. Schematic symbols for depletion-mode MOSFET's. At A, single-gate, n-channel; at B, single-gate, p-channel; at C, dual-gate, n-channel; at D, dual-gate, p-channel.

MOSFET has to literally carve out its own channel! The gate is simply stuck to a section of p-type semiconductor, insulated (in the same fashion as the depletion-mode MOSFET) with a layer of silicon dioxide. The gate current of the enhancement-mode MOSFET, like that of its depletion-mode brother, is essentially zero.

When V_{GS}, the bias between the gate and source, is zero, there is no channel and therefore no current can flow from the source to the drain. If we apply a positive voltage to the gate electrode, some of the p-type material near the gate changes to n-type material. This creates a channel, and current can flow. The greater the positive voltage we apply to the gate, the wider this enhancement region becomes. There is a limit to how well the device will conduct, however; once the channel reaches a certain width, further positive bias on the gate will have little effect. We call this the saturation point. Negative voltages on the gate will not make anything happen at all; the MOSFET will refuse to conduct. Figure 9-12 shows the schematic symbols for enhancement-mode MOSFETs. Like all other semiconductor devices, we may interchange the p-type and n-type materials and the polarity, and still have a workable device. It's nice to have a choice!

There certainly are a lot of different types of field-effect transistors! It's hard to keep them straight. Figure 9-13 should be of

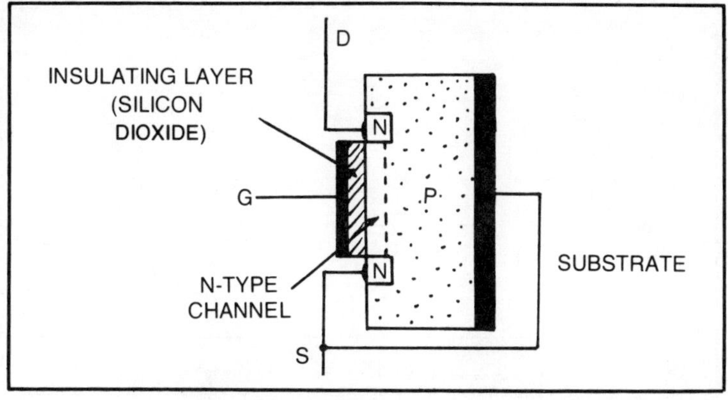

Fig. 9-11. Cross section of an enhancement-mode MOSFET. When a positive voltage is applied to the gate electrode, part of the p-type material reverses and becomes n-type. This creates a channel. The device shown here is an n-channel enhancement-mode MOSFET; reversing the materials results in a p-channel device.

help here; all the various kinds of FETs are categorized and subcategorized in this hierarchial block diagram.

THE FET AMPLIFIER

The n-channel JFET is biased just like a triode vacuum tube, except that the voltages are smaller—in the range of 10 volts between the source and drain, as compared to hundreds of volts between the cathode and plate of a tube. A typical amplifier circuit arrangement is shown in Fig. 9-14. This circuit is ideal for weak-signal amplification in the rf range. The resistance between the gate and ground is extremely high because the gate-source pn junction is reverse biased. (Although there is no separate battery to provide this bias, the resistance of the channel causes a potential difference to exist inside the JFET as reverse bias.)

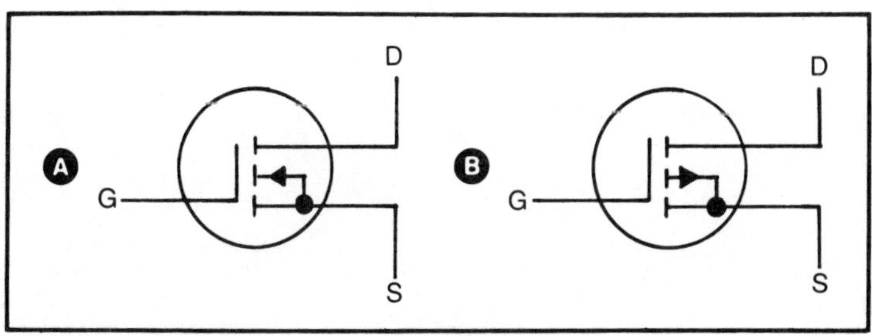

Fig. 9-12. Schematic symbols for enhancement-mode MOSFET's. At A, n-channel; at B, p-channel.

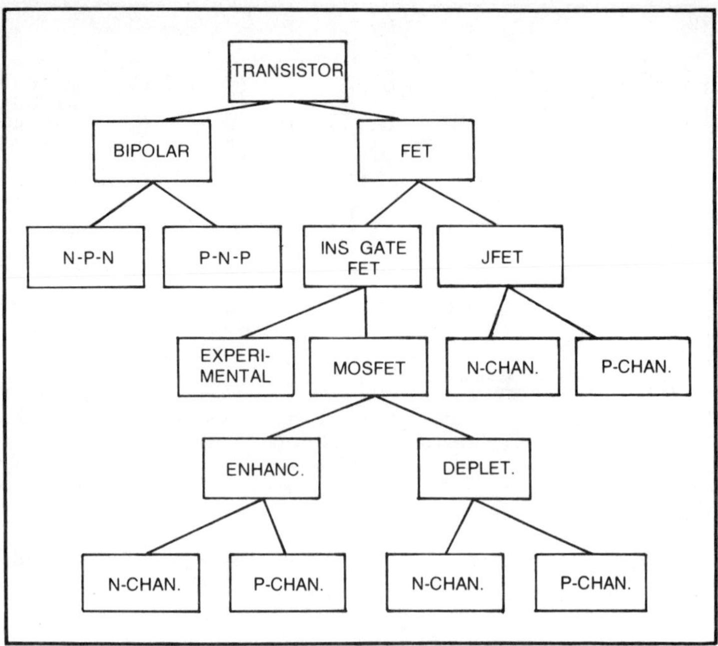

Fig. 9-13. Transistors may be categorized according to a scheme such as this. Dual-gate devices are not shown. Such is the great variety of transistor types.

The way we have set up the JFET in the circuit of Fig. 9-14, the tiniest voltage fluctuation at the gate will result in a considerable change in the channel resistance. In turn, the drain current I_D varies. The more negative the voltage at the gate becomes, the smaller the

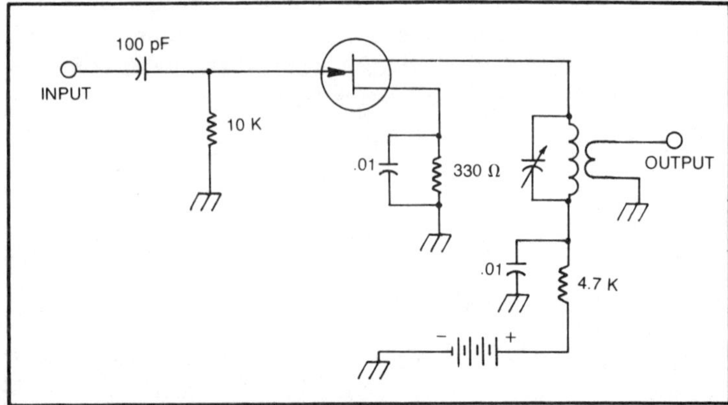

Fig. 9-14. Simple rf amplifier circuit using an n-channel JFET. This is a common-source amplifier, since the source is at rf ground.

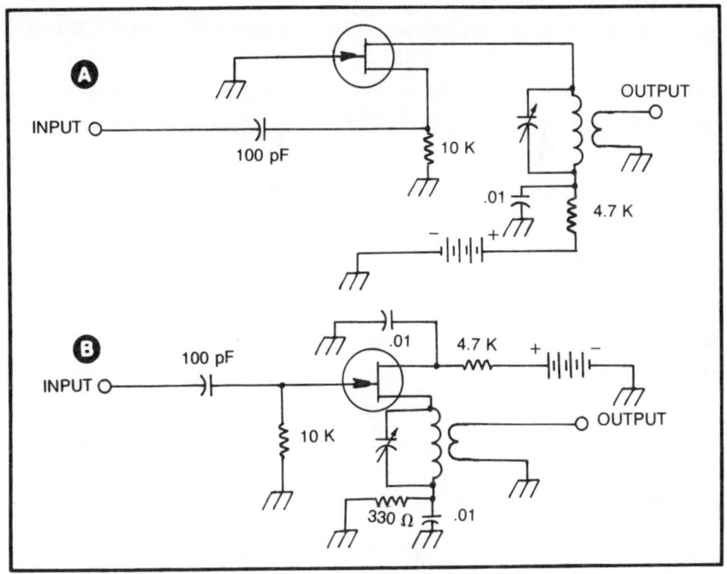

Fig. 9-15. Common-gate (A) rf amplifier, and common-drain (B) rf amplfifer using an n-channel JFET.

drain current I_D, and the more positive the drain voltage becomes with respect to ground. The less negative V_{GS} gets, the more drain current will flow, and the less positive the drain voltage. Therefore, the JFET, in this circuit, reverses the phase of a signal passing through it. We call this a common-source or grounded-source amplifier, since the source is kept at rf ground potential. This is similar to the grounded-emitter circuit of a bipolar transistor.

If you think we can wire a JFET in common-gate and common-drain fashion, imitating the common-base and common-collector bipolar circuits, you're right! Such circuits are also the equivalent of the common-grid and common-plate tube circuits. Figure 9-15A illustrates a typical n-channel JFET amplifier in common-gate arrangement. Figure 9-15B shows a common-drain amplifier. A common-gate amplifier has a low input impedance and thus draws current from the input signal. It is an excellent choice for rf power amplifiers, but doesn't work out too well for weak-signal applications. (A FET as a power amplifier? Yes! Power FETs do exist. Any FET can be used to provide a half watt or so, and we'll look at FETs for higher power levels shortly.) The input and output signals of the common-gate amplifier are in phase. The more negative the source voltage, the more drain current will flow, and the less positive the drain potential.

The common-drain circuit, also known as a source follower, has high input impedance (just like the common-source circuit) but low output impedance. The voltage gain of this amplifier is always somewhat less than 1, while the other amplifiers have voltage gains greater than 1. What, then, is the source follower good for? Well, it functions very well as a broadband amplifier for impedance matching.

At this point, let's sidestep for a moment. We have spoken about voltage gain and current gain; all semiconductor amplifiers have one or the other, and some are designed to have power gain. We have spoken of a given amplifier as having a voltage gain of, say, 10, meaning that its output signal voltage is 10 times its input voltage. But if you're going to talk much about signal gain, you ought to be familiar with the way it is usually expressed: the decibel.

DEFINING GAIN

Suppose we have an amplifier whose input signal voltage is E_{in} and whose output signal voltage is E_{out}. The gain in decibels, abbreviated dB, is simply

$$\text{gain (dB)} = 20 \log_{10}(E_{out}/E_{in})$$

That may look pretty bad, but actually it's very simple. If $E_{out} = 10$ mV and $E_{in} = 1$ mV, then the gain is 10. But in dB, we express that as

$$\text{gain (dB)} \begin{array}{l} = 20 \log_{10}(10/1) \\ = 20 \log_{10}(10) = 20 \times 1 = 20 \end{array}$$

The same formula holds for determining current gain. If an amplifier has an input current (from the signal) of I_{in} and an output current of I_{out}, then the gain in dB is

$$\text{gain (dB)} = 20 \log_{10}(I_{out}/I_{in})$$

For power, the formula is a bit different. If the signal input power is P_{in} and the output power is P_{out}, then the power gain is

$$\text{gain (dB)} = 10 \log_{10}(P_{out}/P_{in})$$

A circuit may not always have gain. We've just seen one example of this: the source follower. This circuit shows a voltage loss. Imagine that we put a 200 mV signal into a source follower and get only 100 mV out. Then the voltage gain is

$$\text{gain (dB)} = 20 \log_{10}(100/200)$$
$$= 20 \log_{10}(0.5) = 20 \times -0.3 = -6$$

A loss is expressed, then, as a negative gain. We may say that the above circuit has −6 dB voltage gain, or we may say that it has 6 dB voltage loss.

Typical JFET amplifiers of the common-source configuration show voltage gains from 10 to 20 dB, with 15 dB probably being about average. Bipolar amplifiers are about the same. The JFET is often preferred for weak-signal amplification, because there is essentially no current drawn from the input of a common-source JFET circuit. This results in high impedance and a low level of electrical noise: Not as many electrons are flying around in the input pn junction as is the case with a bipolar transistor! Lower noise, but the same gain, means a more sensitive receiver. We'll have more to say about noise a bit later in this chapter.

DUAL-GATE MOSFET AMPLIFIERS, MIXERS, AND MODULATORS

There are ways of getting even more than 20 dB gain from a single-stage amplifier. Dual-gate MOSFETs allow us to obtain as much as 25 dB voltage gain, and perhaps even 30 dB at low frequencies. A dual-gate MOSFET amplifier circuit is shown in the schematic diagram of Fig. 9-16.

The input to the amplifier is applied to the signal gate, gate 1, and a fixed positive dc voltage is applied to the control gate, gate 2. While gate 1 operates in the depletion mode because of the distance between it and the source, gate 2 operates in the enhancement mode because of its positive bias with respect to the source. Yes, the dual-gate MOSFET can operate in both modes at once! You might be tempted to think that we could control the gain of this amplifier without running the risk of distortion, simply by varying the bias on gate 2. A more negative potential will indeed reduce the gain of the amplifier, and whatever we do at gate 2 will not affect the characteristics of gate 1. This is quite commonly done; by switching the control gate to ground, we can reduce the gain of this amplifier from about 25 dB to only about 5 dB. This attenuation is useful at times, when very strong signals are present. By putting a potentiometer in the control-gate circuit, we can adjust the gain over a

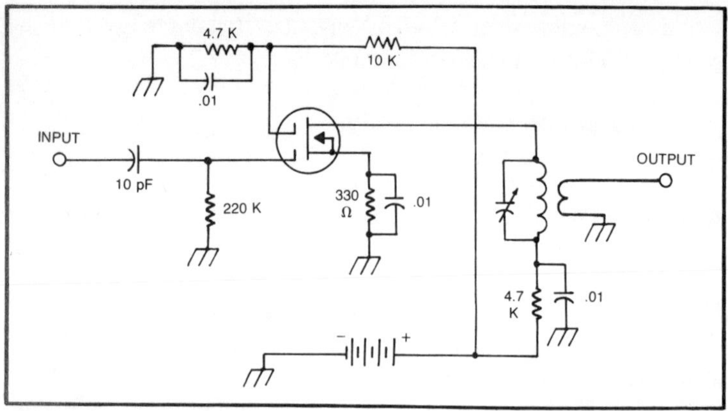

Fig. 9-16. A dual-gate MOSFET amplifier. The signal is applied to gate 1, and a fixed positive bias to gate 2.

continuous range from about 5 dB to 25 dB. Don't try to do this with a bipolar transistor or single-gate JFET or MOSFET! You can't, without changing the operating point of the amplifier, with consequent destruction of dynamic range and linearity.

The dual-gate MOSFET is ideally suited for use as a mixer or modulator. Figure 9-17 shows how we might connect a dual-gate MOSFET in a mixer circuit. We simply apply one signal to gate 1 and the other signal to gate 2! It should remind you old timers of a tetrode tube mixer. Both gates, under the influence of varying signal voltages, affect the drain current I_D, providing sum and

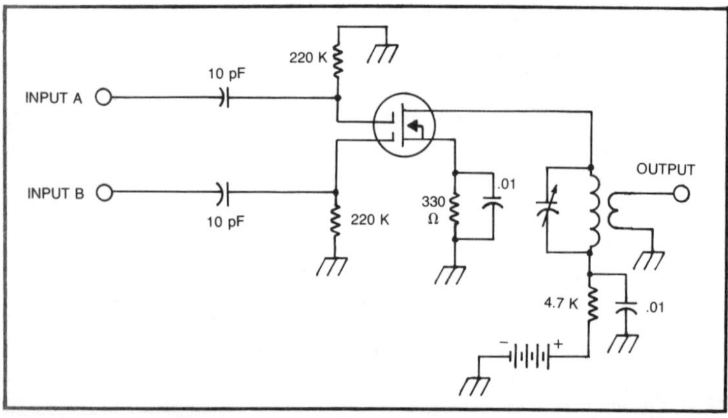

Fig. 9-17. A dual-gate MOSFET mixer. The two signals are simply applied to the two gates. This mixer has much greater isolation between the two input signal sources, and between the inputs and the output, than mixers that employ other devices.

difference frequencies. We may then select between these with a tuned circuit in the output of the mixer.

Now, of course, you may point out that this same thing can be done with a single-gate JFET or MOSFET. And that is true: Both signals can be applied simultaneously to one gate. But using a dual-gate MOSFET provides a tremendous amount of isolation between the two signal sources. The impedance at either gate is so high that the loading effect on oscillators is just about nonexistent. This is because of that good old silicon dioxide layer between the gate electrode and the channel. The dual-gate MOSFET mixer allows much less of the original (and unwanted) frequencies to appear at the output, as compared with mixers designed with bipolar transistors, JFETs, or single-gate MOSFETs.

We have already seen that the AM modulator is very similar to a mixer, and so it should come as no surprise that the dual-gate MOSFET is a natural for this application, too. Figure 9-18 is a schematic diagram of an am modulator that makes use of a dual-gate MOSFET. The carrier signal is applied to gate 1, and the audio is fed into gate 2.

When thinking about the operation of any field-effect transistor, it helps to envision the gates as valves. This analogy is some-

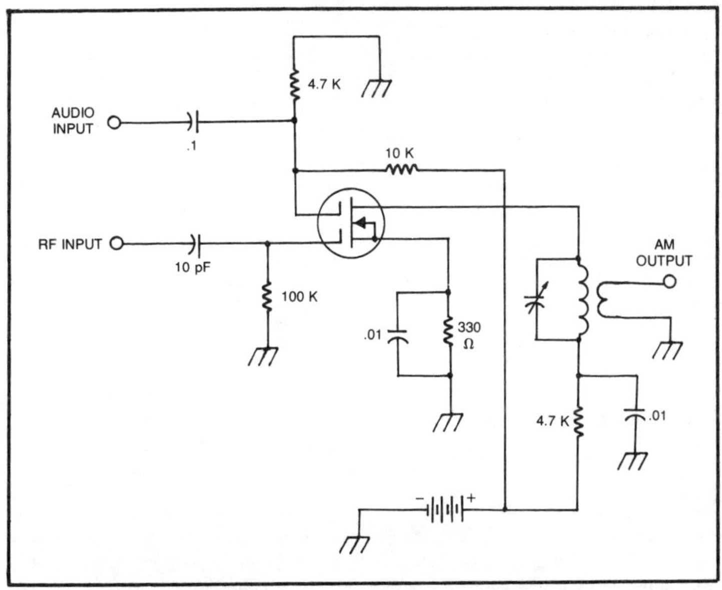

Fig. 9-18. A dual-gate MOSFET AM modulator. The rf signal is applied to gate 1 and the audio to gate 2.

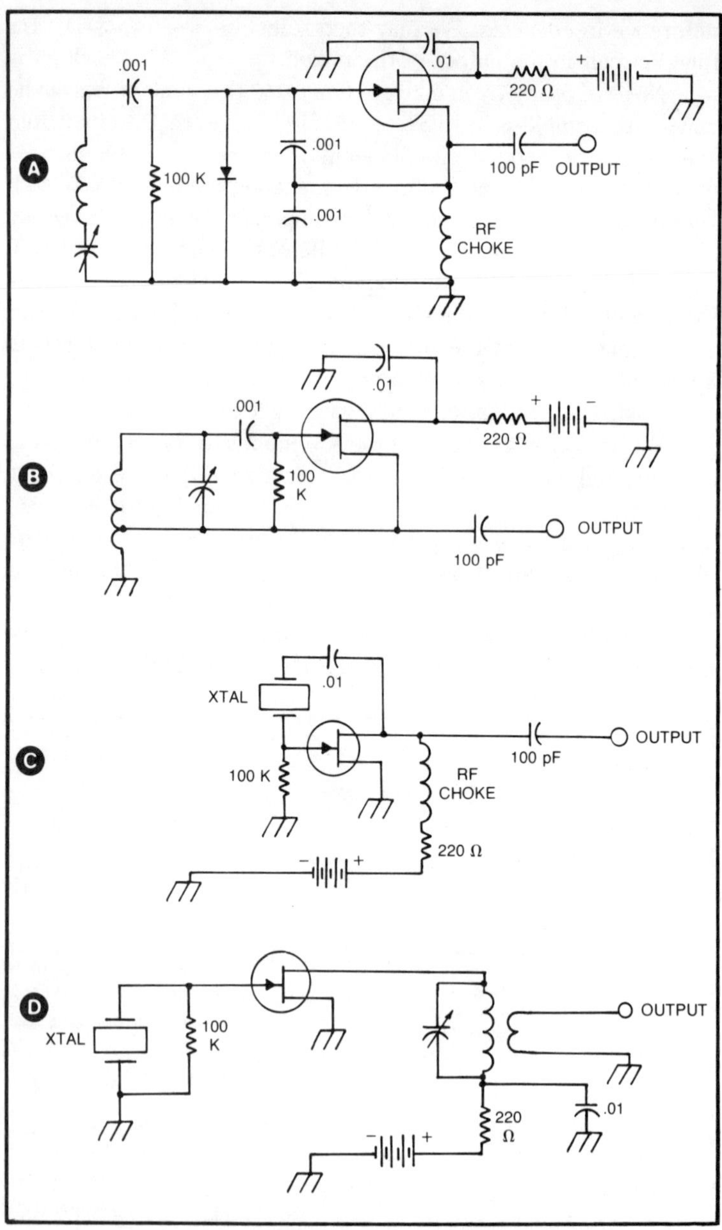

Fig. 9-19. Four oscillators using JFET's. At A, a series-tuned Colpitts oscillator; at B, a Hartley oscillator. These are variable-frequency oscillators. Crystal oscillators are shown at C and D, respectively, Pierce and overtone types. In the overtone oscillator, the tuned circuit is set for an odd harmonic of the crystal fundamental frequency.

what inaccurate in the case of bipolar transistors, but it is very illustrative for the field-effect transistor. In fact, this analogy is even more justified for FETs than it is for vacuum tubes—which in some countries are still called valves! (Don't expect to hear that term often. Transistors have largely taken over.)

Where the vacuum tube and the bipolar transistor have trodden, so goes the FET. We may build oscillators easily using any kind of field-effect transistor. Figure 9-19 shows a selection of FET oscillator circuits. Notice that they are all very similar to their bipolar cousins. Well, an oscillator is nothing more than an amplifier with controlled feedback! So you should certainly not find anything strange about this resemblance.

THE VMOS POWER FET

With all these advantages over bipolar transistors, the FET ought to work well as a power amplifier. Some enhancement-mode MOSFETs are designed especially for use as rf power amplifiers. These devices are known by a variety of names, but probably the most common is vertical metal-oxide semiconductor (VMOS) FET. The VMOS FET hasn't quite caught up with the bipolar transistor in the brute-force department (although it's probably only a matter of time). If you are after hundreds of watts of rf power, you should stick with tubes or bipolar transistors. But for moderate power ranges—about 10 to 30 watts—the VMOS FET is truly the state of the art. VMOS is more stable than the bipolar transistor, and therefore is less likely to break into parasitic oscillation when we really just want it to amplify. The VMOS FET can tolerate a sudden short or open circuit in the output; try that with a bipolar power transistor! Bipolar power transistors are somewhat prone to thermal runaway, a strange form of semiconductor suicide in which increasing heat results in lower efficiency of an amplifier stage, thereby contributing to even more rapid heating, which lowers the efficiency still more, and so on until the device is destroyed. The VMOS FET is more well-tempered. It knows better than to kill itself in this way, even if conditions favor it.

Figure 9-20 is a cross-section illustration of a VMOS FET. There are actually two sources, connected together. This is an n-channel enhancement-mode device. Normally, when the gate is at zero bias, the VMOS FET is pinched off. Applying a positive voltage to the gate causes some of the p-type material to change over to n type. The V in VMOS comes from the observation that the current travels vertically (at least in the illustration). But this is

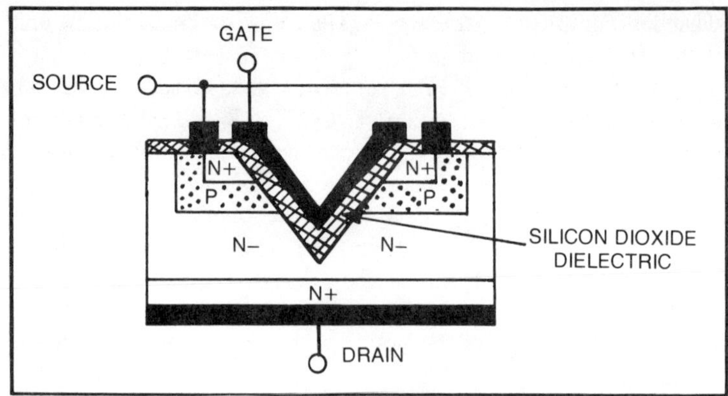

Fig. 9-20. Cross section of a VMOS FET. This is an enhancement-mode device. The current flows in two paths, from each source to the drain. When the gate electrode is positively biased, part of the p-type material reverses to n-type, creating a channel.

really quite arbitrary. We can tilt a VMOS FET any old way, and it will still function, even if the current is going horizontally. But designers of the device had to think of something to call it!

Notice that the VMOS FET has two current paths, rather than one, as is the case with all other semiconductors. This improves the efficiency of the device. A VMOS power FET may be operated as a class-C amplifier with exceptional efficiency, approaching 85 to 90 percent in some cases. The VMOS device performs equally well as a weak-signal class-A amplifier. Talk about versatility! VMOS FETs can be employed in class-A, -AB, -B or -C operation; class C is probably easiest to obtain, since zero bias results in pinchoff.

THE GALLIUM-ARSENIDE FET

All of the transistors and diodes we have seen, up to now, have employed silicon or germanium as the semiconductor material. If you're observant, you have already noticed that there is something noteworthy about these substances: They are elements. The gallium-arsenide field-effect transistor, or GaAsFET , employs a compound as its semiconductor material. Gallium arsenide has better carrier mobility than silicon or germanium. What does that mean? It means the holes and electrons can hop from atom to atom considerably faster. What good is that? Well, carrier mobility is one of the factors that puts an upper limit on the operating frequency of a transistor. The faster the holes and electrons can move, the more cycles per second (hertz) are effectively handled by the device. At uhf and microwave frequencies, the GaAsFET is the amplifying unit of choice.

The GaAsFET is constructed in the same way as the depletion-mode JFET. The gate electrode may be made of aluminum, but better electrical ruggedness is obtained when it is made of gold.

GaAsFETs may be used as oscillators, low-level amplifiers, or power amplifiers at frequencies on the order of 300 megahertz (MHz) to 10 or 15 gigahertz (GHz). (A GHz is one billion hertz.) Like the JFET, the GaAsFET has excellent low-noise characteristics. This is important at frequencies as high as uhf and microwaves, because environmental noise decreases as the frequency goes up, leaving only receiver noise as the limiting factor in sensitivity.

We have talked frequently about the importance of minimizing the noise generated inside a receiver, especially in the front end (first amplifying stage). There is a specific way to evaluate the noise performance of a circuit. If you're at all serious about designing receivers of any kind, you should know about noise figures. Engineers mention this characteristic of an amplifier quite frequently, especially at vhf and above.

THE NOISE FIGURE

Ideally, of course, amplifiers would generate no noise at all. And we all know that such a thing is not possible in the real world. Suppose, for a moment, that it were! Could we then amplify and amplify, using one noise-free stage after another, and have unlimited sensitivity? If this were so, then any signal anywhere in the whole universe would theoretically be audible if we could build enough amplifying stages. But this is not the case. We can't do it, even with perfect amplifiers. Some noise is generated in the environment, because of thermal agitation of molecules. Some noise even comes from stars and galaxies in space. Some comes from the sun. Some comes from the heated mountains of the moon. So the best receiver we could possibly hope for, at any given frequency, is a receiver that doesn't add any more noise on account of its own transistors. We can calculate (but not in this book!) how much noise we could get at the output of an amplifier having a specified amount of signal gain, but no noise generation.

Let's take a specific example. Suppose we have an rf amplifier that delivers 20 dB of power gain. (That's an amplification factor of 100.) If we measure the noise power output of the amplifier, we might get 100 microwatts (uW) with the antenna connected and no signal input. Now engineers can figure out how much of that noise actually comes from the antenna, and how much is generated inside the amplifier. At a frequency of, say, 500 MHz, we might consult an engineer and find that we would get just 50 uW of noise output from an *ideal* amplifier with 20 dB power gain. Let's call the noise output from the ideal amplifier $N_1 = 50$ uW, and the noise from our real receiver $N_2 = 100$ uW. We define the noise figure, F_N, as

$$F_N \text{ (dB)} = 10 \log_{10}(N_2/N_1)$$
$$= 10 \log_{10}(100/50) = 10 \times 0.3 = 3$$

That is, our amplifier has a noise figure of 3 dB at this particular frequency. By convention, all this must take place at an ambient temperature of 290 degrees Kelvin, which is about 63 degrees Fahrenheit.

Another way to define noise figure is to say that F_N is the difference, in decibels, between the signal-to-noise ratio at the antenna and the signal-to-noise ratio at the output of the amplifier. It will always be greater at the antenna, since any amplifier generates noise.

ONWARD

Now we have seen all different kinds of transistors. New ones are still being invented. Someday, perhaps transistors will take over every last bastion of the tube domain! They have just about done it already, but super-high-power amplifiers must still use tubes. This is the next major challenge for the transistor.

Let's look at some more of the ways that transistors, both bipolar and unijunction, are employed in the design and construction of practical circuits.

QUESTIONS

1. Why is a field-effect transistor called a unijunction device?

2. What is pinchoff? What two things can cause it?

3. How does the input resistance of a FET compare with that of a bipolar device, assuming both are properly biased?

4. What semiconductor is used in the GaAsFET? Why is it

preferable to silicon or germanium at uhf and microwave frequencies?

5. What is the difference between depletion mode and enhancement mode in a MOSFET?

6. How would the gate of a depletion-mode MOSFET be biased for class-A operation? How would the gate of an enhancement-mode MOSFET be biased for class-A operation? Assume both devices are n-channel types.

7. What are the relative input and output impedances of the source follower?

8. If the voltage gain of an amplifier is 25 dB and the input signal measures 1 μV peak, what is the peak voltage of the output signal?

9. What advantage does a dual-gate MOSFET offer as a mixer?

10. What are some operating advantages of the VMOS FET over bipolar transistors? What is the chief limitation of the VMOS FET?

Chapter 10

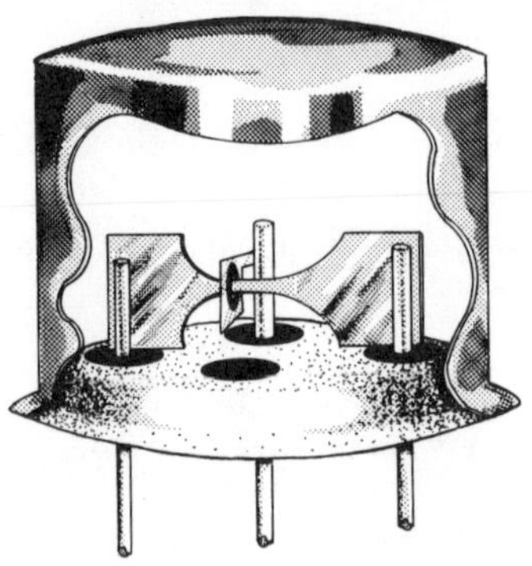

Practical Circuits

IT MAY SEEM VERY ODD THAT WE SHOULD HAVE A CHAPTER ENTItled Practical Circuits. And we wouldn't be at all surprised if you were to demand, "And what have we been doing up to now? Studying medicine? Law? Business?" True! Quite true! Take a single component such as a resistor, connect another component or a battery, and we have a circuit. We've been doing that almost from the very beginning of the book. But what sort of circuits have you and we gone through together? Wouldn't you say that they've been the common-garden variety, the sort of stuff you would logically expect? We've had audio and rf amplifiers, converters, mixers, and oscillators. And before that we had a generalized conglomeration, just so that we would get the idea of how transistors behave in the polite society of resistors, capacitors, and batteries.

If we were to stop right there you would have every right to say, "So what? Is that all transistors can do?" Quite correctly you expect more, so let's examine some other transistor accomplishments. Before we do, though, we are sure you know by now that a transistor isn't just a vacuum-tube substitute. We can't take a vacuum-tube circuit (remove Dr. de Forest's brainchild), put in a transistor, and then say, "There, that's that." Hardly! Transistors have their own ideas as to the components that should accompany them. Does this mean, then, that you are going to need to learn a whole line of new circuits and that your knowledge of vacuum-tube circuits has gone down the drain? No—for you will see that the basic

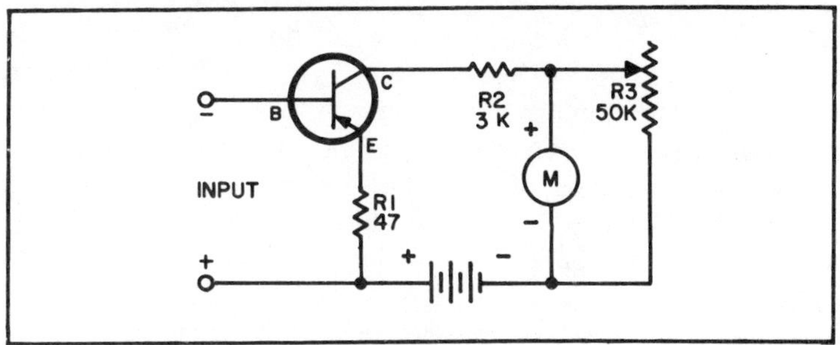

Fig. 10-1. Insensitive meter can be used to measure currents in the order of microamperes with the help of a transistor amplifier.

circuit idea has remained the same and in many cases you will be able to identify the circuit from earlier experience and knowledge. So let's start in an easy way and see what we can do together. The purpose here is to learn more about how transistor circuits work through theory analysis. We'll finish up with a construction project.

METER AMPLIFIER

Of all dc milliammeters ever manufactured, the one having a full-scale deflection of one milliampere (sensitivity, 1,000 ohms-per-volt) was probably the most common. At one time this meter was a workhorse but with improvements in magnets and manufacturing techniques, much more sensitive meters are being used. The great big hitch here, of course, is that the more sensitive the meter, the more it costs. The circuit shown in Fig. 10-1 will give a meter an apparent ten-time increase in sensitivity, but by the time you get through paying for the parts, you may not have made any savings at all. But at least its nice to know that you have a choice. And there's a limit to how sensitive a meter instrument can be made! A circuit such as this is thus very useful when we need a super-sensitive meter.

To understand the circuit of Fig. 10-1, think of the current amplification capability of a transistor. In an earlier chapter we called this beta. It is the ratio of the change in collector current to a change in base current. For a representative transistor, beta could have a value, say, of 50. Thus, if we had 20 microamperes of base current, we would have 1,000 microamperes (or a milliampere) of collector current. And if the base current changes, we get 50 times as much change in the collector current.

Now suppose that in the circuit of Fig. 10-1 we had a dc milliammeter in the base circuit of the transistor. We would need a very sensitive meter indeed to give us a full-scale deflection of 20 microamperes. But if we had a dc meter in the collector circuit, we could easily use a 1 milliampere movement. We could calibrate this meter in microamperes and when it would read full-scale we would be measuring a current of only 20 microamperes.

Is the circuit new? Not actually, since it is a very ordinary, very simple transistor amplifier. What we have here is a dc amplifier with but one new component. This is potentiometer R3, a calibration adjustment that is varied until the meter reads full-scale.

It is important to understand just what it is that the meter in the collector circuit does. It just reads collector current. But this collector current is the same as the base current multiplied by the beta of the transistor. Thus we can take advantage of the amplifying action of the transistor to measure small currents. Remember, though, that these small currents are not measured directly, but indirectly.

RELAY AMPLIFIER

Just as the circuit in Fig. 10-1 permits us to use a less sensitive meter (and therefore a less expensive meter) so too can we use this circuit, or some variation of it, to give us an apparent increase in the sensitivity of other components. A relay is one such item. Relays, like meters, are assessed in terms of their sensitivities.

Figure 10-2 shows the relay circuit. The relay might require a current of one or more milliamperes to attract the armature of the relay toward its core. In the circuit of Fig. 10-2 a current of about 50 microamperes in the input will increase the collector current to the amount required to close the relay. Assuming this current to be 2

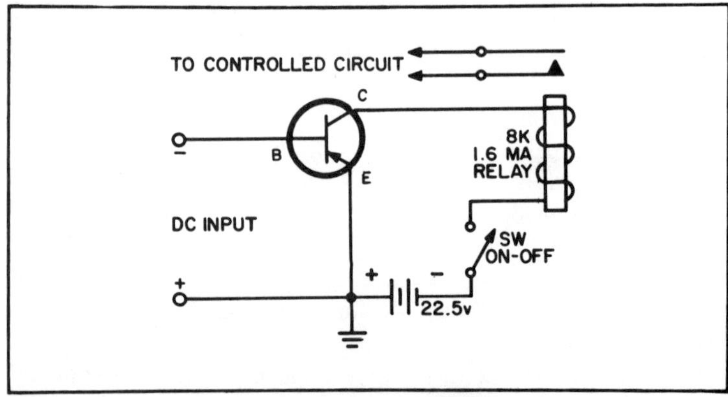

Fig. 10-2. Simple transistor amplifier can be used to operate a relay.

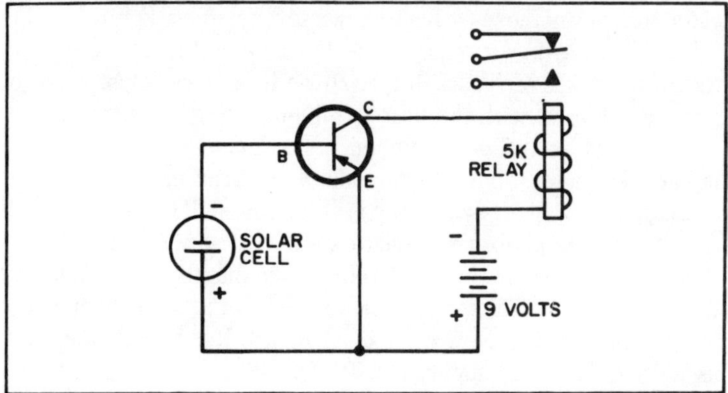

Fig. 10-3. Relay circuit will operate when solar cell receives light.

milliamperes (or 2,000 microamperes), the gain in our circuit is 40 ($40 \times 50 = 2{,}000$). In this circuit the relay is the load.

In the circuits of Figs. 10-1 and 10-2, what would be our input? It could be some sort of signal which would give us the necessary amount of forward bias to activate the collector circuit. The input current could also be produced by a solar cell, as shown in Fig. 10-3. Now take a few moments out to note the considerable similarity these circuits have.

PREAMPLIFIER CIRCUIT

At first glance, the circuit in Fig. 10-4 looks far more compli-

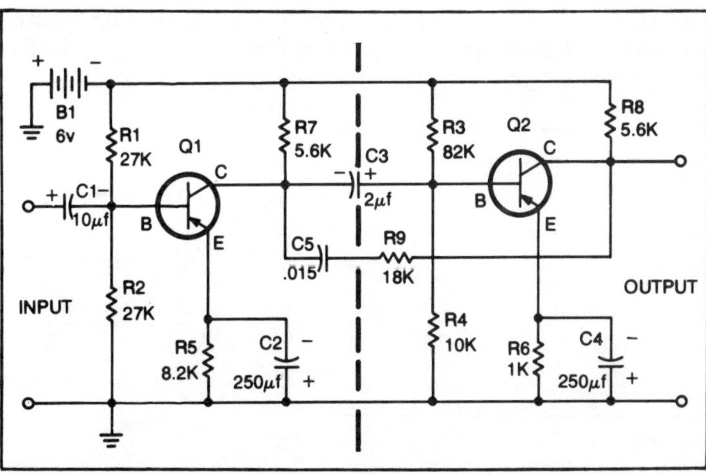

Fig. 10-4. Resistance-coupled audio amplifier consists of a pair of amplifiers, one driving the other.

cated than any of those we have shown so far in this chapter. For this reason we have drawn a dashed line down the center of the circuit so that you can see for yourself that all we have here is just a pair of audio amplifiers, with the first amplifier driving the second.

We have analyzed the components in circuits similar to this one in earlier chapters, but let's take out a few brief moments to do it again, just by way of review. If you will examine R1 and R2 you will see at once that these two resistors are in series. If you will trace their connections to battery B1 you will see that they are connected directly across it. These two resistors supply the correct amount of forward bias for Q1. Now what about R3 and R4? Exactly the same as R1 and R2! They supply bias for Q2.

Note that R1 and R2 do not have the same values as R3 and R4. If Q1 and Q2 are different you would expect their forward biasing requirements to be different also. The two transistors also use emitter bias. The emitter resistors, R5 and R6 are shunted by large values of capacitance—250 μF. These capacitors have a rating of 6 volts. While we're on the subject of capacitors, examine C1 and C3. These are also fairly large capacitance units (compared to the types used in vacuum-tube amplifiers) and are 25-volt electrolytics. We must watch our polarity in the case of electrolytics. When considering the emitter bypass capacitors, remember that for pnp transistors the electrons flow from the emitter down to ground. This makes the top end of the emitter resistor minus and the bottom end plus. C2 and C4 are connected to observe this polarity.

In the case of C2 and C4 the matter of polarity is quite clear cut, but what do we do about coupling capacitors C1 and C3? It is entirely possible for a coupling unit to be connected to two positive voltage points, yet we must still be careful about polarity.

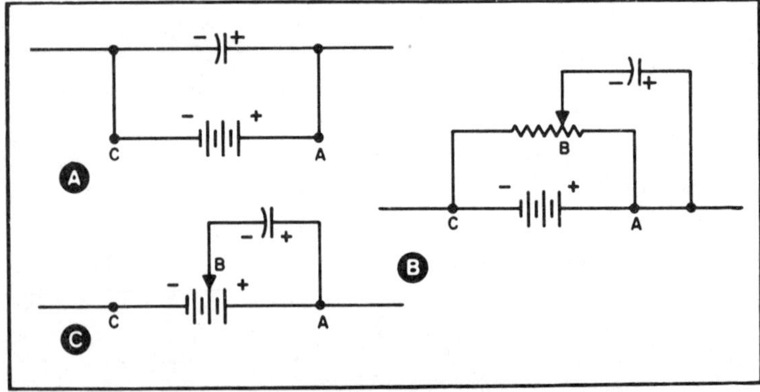

Fig. 10-5. Methods of connecting an electrolytic capacitor.

To see how we can come to a decision about this, please look at Fig. 10-5. In Fig. 10-5A we have an electrolytic directly across a battery. There is no problem here since the plus terminal of the electrolytic goes directly to the plus terminal of the battery. Similarly, the negative lead of the electrolytic is wired to the negative end of the battery. The electrolytic receives the full voltage of the battery and must be able to withstand it.

In Fig. 10-5B, we are still observing polarity since the negative end of the electrolytic is connected to the arm of the potentiometer. But isn't point B a positive point? Yes, it is, but it is less positive than point A.

To simplify matters even further, we have eliminated the potentiometer in Fig. 10-5C. Now you can see that we have connected the negative terminal of the capacitor to one of the positive terminals of the battery. This is point B. But so is point A. However, while point B is positive, it is less positive than point A. Another way of saying exactly the same thing is to say that point B is negative with respect to point A.

Now getting back to Fig. 10-4, how do we know that we have connected C1 and C3 correctly, so that polarity is observed? In the case of C1 we don't know. We have not been told just what it is that C1 connects to. It is possible that in connecting C1 to its input we would find it necessary to turn C1 around.

We have no such doubts about C3. The collector of Q1 to which this capacitor is connected is more negative than the base of Q2, to which the capacitor is also connected.

R7 and R8 are easy to identify. These are our collector load resistors. What is left? Just a series combination of C5 and R9. This is our negative feedback circuit. It improves the stability of the amplifier and gives the amplifier a somewhat better frequency response.

We have gone through this amplifier rather hurriedly, since audio amplifiers were covered in an earlier chapter. A simple resistance-capacitance coupled amplifier of this sort could be used as an audio preamplifier.

SIGNAL TRACER

A signal tracer is just an audio amplifier plus a detector. The detector is generally a crystal diode as shown in Fig. 10-6. For the amplifier portion we could have used the circuit of Fig. 10-4, but we have added the one in Fig. 10-6 just to emphasize the fact that not all audio amplifiers are exactly alike. As an example, locate C3 (Fig.

10-6). If you will trace its connections you will see that it is shunted directly across the battery. The purpose of this capacitor is to maintain a low-impedance path around the battery. As long as the battery is fresh, or at least not too used, its impedance will be fairly low. But with time and use the internal resistance of the battery will rise, and long before a decision is made to discard it, will act as a common impedance or coupling device between stages. This can result in feedback between stages, producing amplifier instability. An electrolytic across the battery ensures a low-impedance path around it regardless of the condition of the battery. But before we decide that this is a wonderful cure all, remember that electrolytics also get old and that they have their own ills—including an increase in their own internal impedance.

Now examine the coupling capacitors in Figs. 10-4 and 10-6. How do they compare as far as capacitance and polarity are concerned? C3 in Fig. 10-4 has the same capacitance and the same positioning as coupling capacitors C2 and C4 in Fig. 10-6. But what about the input capacitors (C1 in Fig. 10-4 and C2 in Fig. 10-6)? Quite a difference here in several respects. Note the arrow in Fig. 10-6 showing the direction of current flow through the diode load resistor R2. When the slide arm is at the bottom end of the resistor the negative end of C2 is connected to the negative end of the battery. But suppose the slide arm is at the top end of the potentiometer. The lead of C2 is now at some plus point depending on the amount of current being rectified by the diode and the value of the gain control. This might be as much as 1 or 2 volts. Now what about the other side of C2? This side of the capacitor is connected to the base of Q1. From the base we have two ways of approaching the

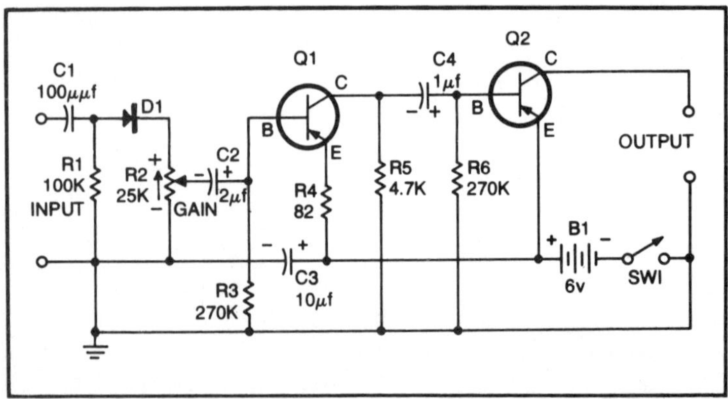

Fig. 10-6. The diode permits use of the audio amplifier as a signal tracer.

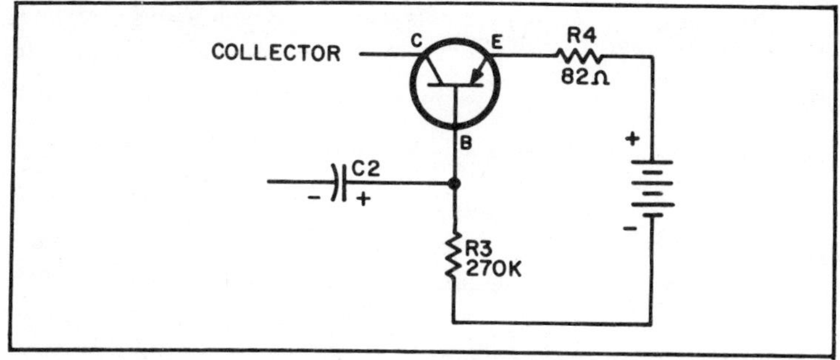

Fig. 10-7. Rearrangement of part of the circuit of Fig. 10-6 to show voltage-divider action.

battery. One path is through a fairly high value of resistance (R3—270,000 ohms) to the negative terminal of the battery. The other path is through the transistor to the emitter, through 82-ohm emitter resistor R4 to the plus terminal of the battery. This latter path is so very much lower in resistance than R3 that the base (hence capacitor C2) is not too far removed (in terms of voltage) from the plus terminal of the battery.

Since this may not be too clear from an examination of the circuit, we have re-drawn it for you in Fig. 10-7. Consider the emitter and base of the transistor as a resistor of low-value in series with R4 and R3. These three components act as a voltage divider across the battery. But because of the very large value of R3, the plus connection of C2 is much closer, from a voltage viewpoint, to the plus terminal of the battery.

While we are making comparisons between the two amplifiers, is there anything else we should notice? Yes—there are several items we should spot. Q2 in Fig. 10-6 has no emitter resistor. The emitter resistor for Q1 isn't bypassed. Also, the two transistors, Q1 and Q2 act as part of a voltage dividing network. In Fig. 10-4 however, we had an actual resistive voltage divider network (R1 and R2; R3 and R4) for forward bias. This arrangement is better than compelling the transistor to work in a resistive function. Also, in Fig. 10-6 we have no feedback network. From all this you might conclude that Fig. 10-4 is the better circuit, and so it is. But before we toss out Fig. 10-6, consider the uses of the two circuits. One is an audio preamplifier where we have some interest in stability and sound quality. The other is intended as a rather ordinary audio amplifier to be used as part of a test instrument.

It is possible that the difference between these two circuits may have been immediately apparent to you. Carry this sort of analysis along to transistor radio receivers and you will at least have learned a few of the reasons why there is a price differential among them.

ANOTHER AUDIO AMPLIFIER

As long as we're examining audio amplifiers, let's take a look at the one shown in Fig. 10-8. This is a transformer-coupled unit, but a first glance seems to show some component arrangements worth investigating. The emitter circuit for Q1 looks a bit odd until we examine it more closely and see that some of the components have been drawn in such a way as to make the circuit look peculiar. R4 and C3 we can recognize immediately as the emitter-resistor and bypass capacitor. But what about R2, R3, and C5. We can't do much about these until we grope around a bit and locate R8. Look just a bit more carefully and you will see that R2, R3, and R8 are all in series and that this series network is connected directly across the battery. What is this network? It supplies forward bias for Q1. Similarly, R5, R6, and R8 have the job of forward biasing Q2.

What other information can we extract from this circuit? C4 in the output circuit of Q1 might require a bit of thinking. The small value of C4 indicates that it is an rf bypass, somewhat unusual to find in an audio amplifier. However, if the input is supplied through a diode detector, we might find C4 necessary. It is possible, however, that the diode detector might have an rf bypass in its own output circuit, so C4 might be an unnecessary luxury. C7, in the output of Q2 is a fixed tone-control capacitor.

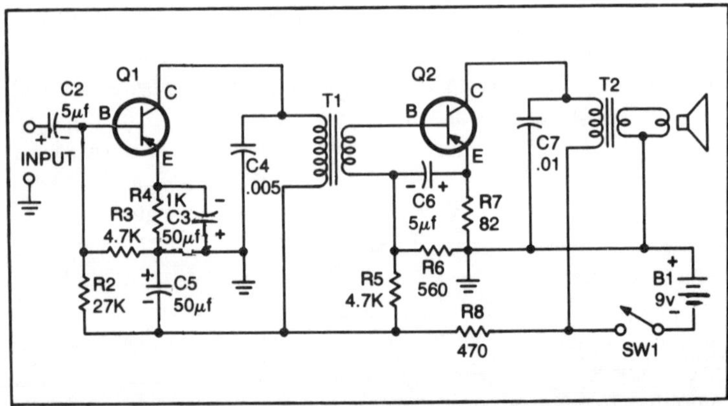

Fig. 10-8. Sometimes radio components are drawn so that their function isn't too obvious. When this happens start tracing the circuit and all the parts will fall into their proper places.

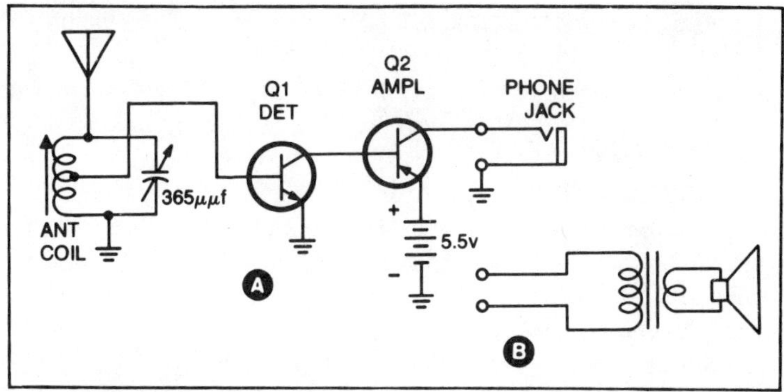

Fig. 10-9. Transistor radio receiver with a minimum number of parts.

LOOK MA—NO PARTS!

The circuits we have examined so far in this chapter are certainly modest enough in their demands for parts, yet you would have a complete transistor receiver with still fewer components. We have the circuit in Fig. 10-9. And just in case you think that this is good for earphone reception only, we have included provisions for a speaker.

A receiver of this sort will not have the sensitivity or selectivity of a six-transistor set, nor is there any claim of this sort being made. What we are interested in here is not in running a contest between receivers but in learning more about the functioning of transistor circuits.

With this caution in mind, what is there about Fig. 10-9 that can add to our storehouse of knowledge? You might think that the obvious absence of parts might clarify the situation.

The first item we might notice is that Q1 is being used as a detector. The fact that it is marked detector is a sure-fire giveaway. How could we know, then, that it is a detector without the help of the label? There are several clues that demand attention.

Q1 is located between a transistor (Q2) being used as an audio amplifier and a tuning coil. Even if Q2 were not marked, we see that it is connected to a phone jack. Consequently we can draw the conclusion that the input to Q2 must be audio. And since the input is a tuned rf circuit (it would have to be with a 365 $\mu\mu$F tuning capacitor) then the only function left to Q1 would be detection.

Suppose, though, we assume this isn't enough evidence. What more information do we have? What about the forward bias for Q1? Actually, there is no forward bias other than that supplied by the input signal. And since this is the case, we may assume that the operating point on the characteristic curve must be in the vicinity of the cutoff point. In this circuit we depend on one-half of the input signal to supply enough forward bias to take the transistor out of cutoff. The other half of the signal will drive it further into the cutoff region. This is rectification, or, to give it its other name, detection.

What about Q2? Is it biased at all? Is it properly biased? Since we have a pnp transistor we want our emitter to be positive with respect to the base. We can see quite readily that the emitter connects to the plus terminal of the battery, but what about the base. The return path of the minus terminal of the battery is through the emitter-collector circuit of Q1. Before we decide this is not for us, consider the emitter-collector of Q1 as a resistive path and you'll have no trouble getting back to the negative terminal of the battery.

Working with crystalline materials, as we are, does not necessarily imply that all our explanations are going to be crystal clear. Earlier in this chapter we mentioned that a circuit diagram will often obscure an understanding of the theory. For this reason let us consider the biasing of Q2 once again, but this time from the vantage point of Figs. 10-10A and B. In drawing A the biasing arrangement for the transistor is the same circuit arrangement we have used so often in earlier chapters. But what about drawing B? We don't have to connect the base of Q2 into the divider since Q2 and Q1 in series act very much like the resistive pair shown in Fig. 10-10A.

COMMON-COLLECTOR AMPLIFIER

We guarantee one thing. Once you get accustomed to the idea of circuit analysis, you will not be satisfied to look at a circuit without really seeing it as it is. Circuit analysis is guaranteed to keep your mental gear-box turning, and that is as it should be.

So, flexing our gray-matter muscles, we come across Fig.

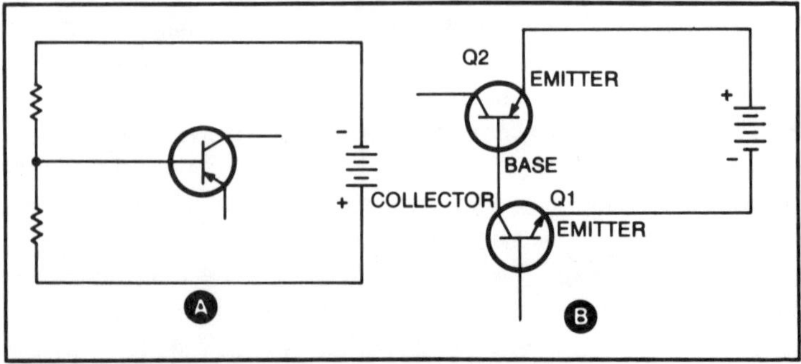

Fig. 10-10. A pair of transistors can be used as a voltage divider.

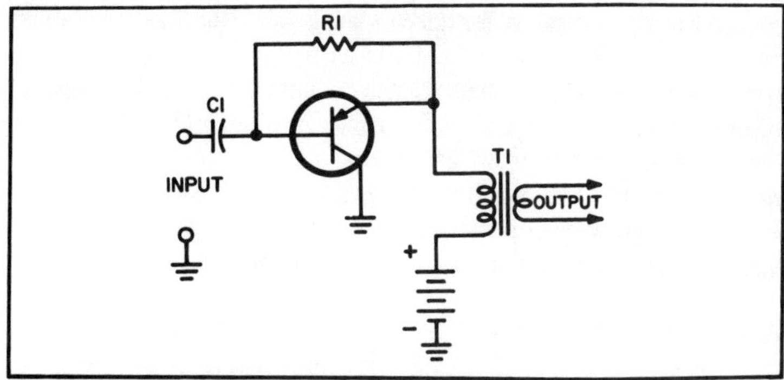

Fig. 10-11. Grounded-collector amplifier. The true relationship of R1 to the transistor isn't clear until the circuit is redrawn.

10-11, which does not deceive us for a moment because of its simplicity. We recognize it immediately as a grounded-collector amplifier, but classifying the circuit is just a small step. There aren't many components, so we can dispose of them fairly rapidly. C1 is a coupling capacitor connecting the circuit to a preceding stage or some signal source. Ditto for T1 except that it couples the output to the next stage.

Since R1 is the only component remaining, we might become sort of complacent and think that we have analyzed the circuit. But have we? How is the circuit biased? And just what is R1 up to anyway? Since R1 is obviously involved in the biasing, answering one question will probably give us the solution to both.

Since this is a pnp transistor, we want a positive emitter and that is exactly what we do have. The emitter is connected to the positive terminal of the battery through the primary of T1. For this type of transistor we want the collector negative, and since both the collector and the minus terminal of the battery are grounded, that requirement is filled.

R1 is still dangling in mid-air and so is the question of how we get our forward biasing. The way the circuit of Fig. 10-11 is arranged isn't too much help in supplying an answer, so let us rearrange the circuit as shown in Fig. 10-12. Once again we have a voltage divider, but this time it consists of the primary of the output transformer in series with R1 and the base-collector circuit of the transistor. Think of the transformer primary winding as a resistor and the base-collector circuit of the transistor in the same light and you will see the voltage divider for what it really is. And here, just as in Fig. 10-10A, the base is connected directly into the voltage divider for proper forward biasing.

What do we need for circuit analysis? Our first and most obvious need is to toss out complacency and for it substitute the realization that a circuit diagram is a convenience and as such it can often obscure as much as it explains. The second and not such obvious need is to re-draw that portion of the circuit diagram that is not absolutely clear so that its operation and function become evident. This isn't an operation that will appeal to lazy people, so, if you wish, you can use it as a measure of character analysis.

OSCILLATORS

Many people like to classify their ideas and their thoughts, and by putting them into compartments in their mind are able to keep all their facts nice and orderly. If we go along with this sort of thinking, a logical step would be to put all transistor circuits into two categories—amplifiers and oscillators. What we have talked about in this chapter are amplifiers, but it isn't such a big jump to oscillators as you might think. Some amplifiers will oscillate at the slightest opportunity and frequently we must take steps to make sure that they will not do so. We saw briefly some oscillator circuits in Chapter 7. Let's look at the workings of the oscillator in a bit more detail.

The very close relationship between a multivibrator type of oscillator and amplifier circuitry is very nicely illustrated by the drawing in Fig. 10-13. A quick look might convince you that what we have here is an ordinary resistance-coupled amplifier, but a single component, C2, changes all that. The output at the collector of Q2 is in phase with the base-emitter circuit of Q1. This is the sort of thing

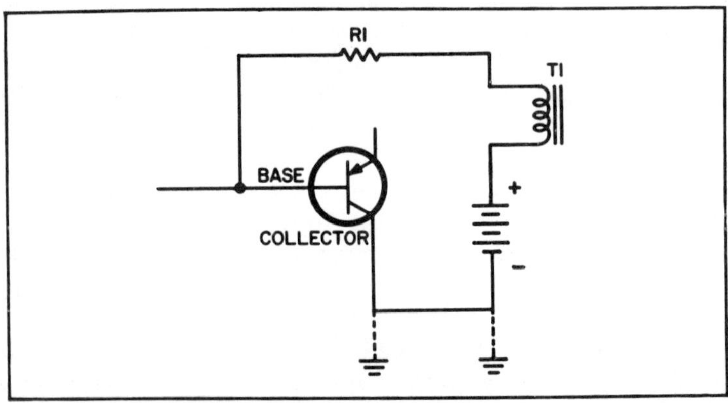

Fig. 10-12. The base and collector of the transistor of Fig. 10-11 are considered here as part of a voltage divider.

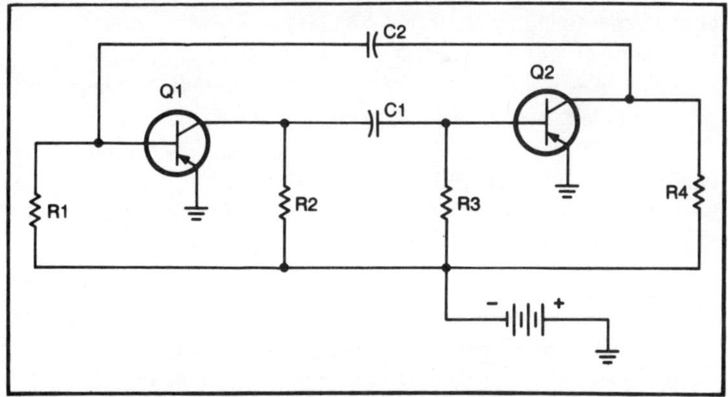

Fig. 10-13. The addition of a feedback capacitor, C2, changes the circuit from a resistance-coupled amplifier to a multivibrator oscillator.

positive feedback is made of, a condition that leads to oscillation. Now you might insist that there is no input to Q1, consequently there can be no signal feedback. It is true we have no-external signal, something we would need if C2 were not present, but consider how unlikely it is that both Q1 and Q2 will conduct exactly the same amount of current at exactly identical moments. Any small voltage change at the collector of Q1 will be sent along to the base input of Q2, through C1. This small voltage change will be amplified by Q2 and fed back to Q1, through C2. What could have caused the current in the Q1 collector circuit to change? This could have been the amplified effect of a very tiny change in the forward biasing of Q1.

It may seem odd to talk about a change in the forward biasing of Q1 since what we have here is a dc circuit. This change, though could consist of a very slight variation in the amount of current flowing between base and emitter. In what direction would this variation be—that is, would the change be such that the base-emitter current increased? Or would it be a decreasing current? it makes no difference. We're not interested right at this point in knowing what the base-emitter current is up to. The important fact is that a change does take place. Please keep in mind, too, that the variation in base-emitter current is practically microscopic—if we can use such a word in connection with current.

That the change is tiny does not bother us either. For this minute change is amplified by Q1, further amplified by Q2 and is then fed back to Q1 (via C2) in such a way that the extremely tiny change receives considerable support. Add the two—the small change and its highly amplified counterpart—and we now have the equivalent of an input signal.

The operation of a multivibrator is simple. Its explanation is quite complex. Briefly, what happens is that each transistor, Q1 and Q2, runs the gamut from cutoff to saturation. The output of this oscillator, when compared to the symmetrical sine waves produced by coil-type oscillators of the Hartley and Colpitts varieties, is highly distorted.

CRYSTAL OSCILLATORS

The silicon and germanium semiconductors we have covered in this book form the very famous and highly-publicized duo of a rather large family of crystalline substances. Other crystals—such as quartz, and tourmaline—have their own, very legitimate claim to fame. These crystals act as true transducers[1] and can change a mechanical strain to a voltage.

The crystal is ordinarily placed between a pair of metal plates. When the plates are connected to a voltage source, the crystal will distort. When the crystal is permitted to resume its natural shape once again, its action produces a voltage or electrical charge across the two metal plates which hem it in. But a crystal can no more resume its exact shape immediately, than a compressed spring can when released. What we will get will be a series of charges on the metal plates, each one smaller than the preceding. Again we can compare this to a spring whose expansions and contractions keep getting smaller and smaller. But we can keep the spring bouncing and we can keep the crystal in action by applying properly timed voltages. In this respect the crystal behaves very much like a coil in series with or in parallel with a capacitor. The coil-capacitor combination will resonate at a particular frequency, depending on its physical dimensions. So will the crystal.

With these facts safely tucked away, let's proceed to an examination of the crystal oscillator circuit of Fig. 10-14. We can recognize some old friends immediately, such as the emitter resistor R1 and its bypass capacitor C1. We have also come across R4, the collector load, in other circuits. There are a few other components whose appearance or function may seem to be new. Start with the crystal. It is represented by a pair of parallel lines used to indicate

[1] A transducer is a substance or device which can change a force, such as torque, pressure, velocity, strain, shear, sound, heat or light, into a voltage, or conversely, can take a voltage and change it into sound, pressure, etc. A microphone is a transducer. So is a speaker.

the metal plates sandwiching the crystal, shown as a rectangle. Shunted across the plates of the crystal is a radio-frequency choke coil, marked rfc. Neglecting the choke coil and the crystal for a moment, what about R2 and R3? These two resistors are in series but, as you can see, they are shunted directly across the battery. The base of the transistor is connected to the junction of these two resistors through the choke coil. The function of the two resistors now becomes evident. Since we have a pnp transistor, we want the base at a slightly lower positive voltage than the emitter and this is just what these two resistors do. Now we can also see the necessity for the radio-frequency choke coil. The dc resistance of this coil is very low, possibly less than 1 ohm, and so the base quite properly considers itself directly connected to the junction of R2 and R3.

The choke coil raises some technical questions in our thinking processes, and if it doesn't, it should. A little earlier we mentioned the action of the crystal in charging the metal plates which hold it. But with a choke coil of just 1 ohm resistance, don't we have what amounts to a short across the plates? Yes, for dc. But the varying charge on the crystal plates is ac and so for this type of voltage the choke coil could have an impedance of thousands of ohms.

How can we regard this circuit? Ignoring the base and the crystal for the moment, current will flow from the negative terminal of the battery, through R4, through the transistor from collector to emitter, through R1 to ground or the positive terminal of the battery. This is a direct current flow and we have done nothing to

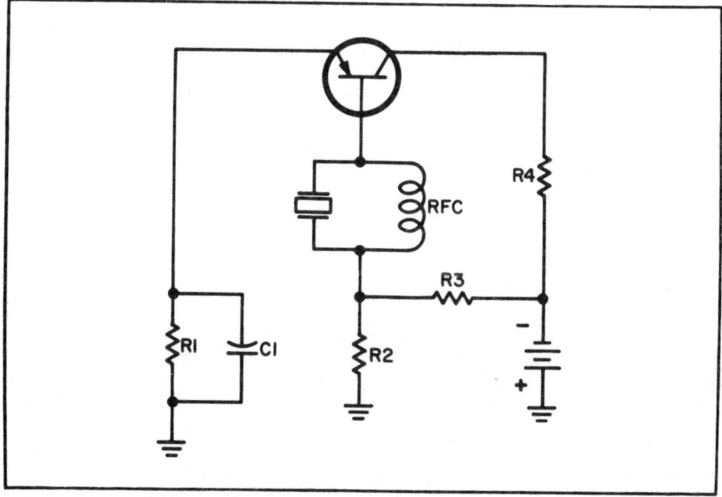

Fig. 10-14. Crystal-controlled transistor oscillator.

change it. But with our natural born penchant for disturbing the existing order you know by now that things are going to change. Connected to the base we have a crystal which will act very much like a tuned circuit. But this tuned circuit, connected as it is between base and emitter, will vary the forward bias. This same crystal is not only in the base-emitter circuit, it is also located in the base-collector circuit. A small part of the energy of the output circuit is fed back to the input, overcoming its losses, and permitting continued oscillation.

MORE ABOUT THE CRYSTAL

An oscillating crystal is a nice challenge to our mental processes because when we get right down to it, the crystal (and its holder) are mechanical. The challenge arises out of the fact that we had better think of it as electrical, if our crystal oscillator circuit is to make any sense. The crystal, though, doesn't hold any monopoly on mechanical devices which can be interpreted electrically or electronically. A stretched spring, released and then allowed to come back to a resting point after its wild back and forth gyrations have subsided, can be represented by a damped wave. The rotating armature of a generator results in a sine wave of constant amplitude.

Figure 10-15 shows the equivalent circuit of the crystal. There is a certain amount of capacitance between the metal plates and, because this shunts the crystal, that is exactly the way in which we show it. The crystal itself behaves like a series-tuned circuit and since every tuned circuit has a certain amount of resistance associated with it, it isn't too surprising to find R in series with L and C. Unfortunately, in a generalized circuit of this form, values of R are not given, and since one unmarked resistor looks very much like another, we have no way of knowing whether R is large or small. In the case of the crystal, more so than in the case of actual tuned coil and capacitor arrangements, R is very low. R, though, influences Q directly, and so the crystal, because of the very low amount of R, has a very high Q.

With this information on hand, circuit analysis of Fig. 10-14 resolves itself very nicely as we see in Fig. 10-16. R, in series with L and C (in Fig. 10-15), has been omitted since there is less need for its inclusion than in an actual coil-capacitor circuit. Since C and L form a series circuit, it will have its minimum impedance at its resonant frequency. At any other frequency than the resonant one, the impedance of L and C rises quite dramatically. The net effect of this electronic behavior is to give us an oscillator that works at a

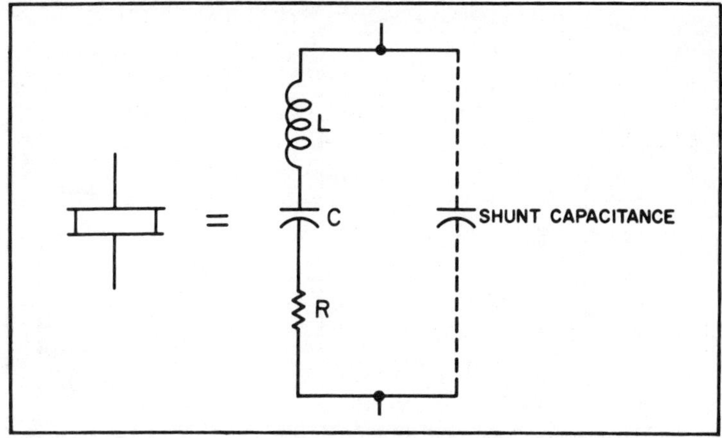

Fig. 10-15. A crystal and its equivalent electrical circuit. The shunt capacitance and the series resistance are small and can often be disregarded.

single frequency, so much so that crystal oscillators are often used as frequency control circuits or where single frequency stability is the prime consideration.

Now what about the choke coil, RFC, in Fig. 10-16? We still need it, for the substitution of the electrical equivalent of the crystal

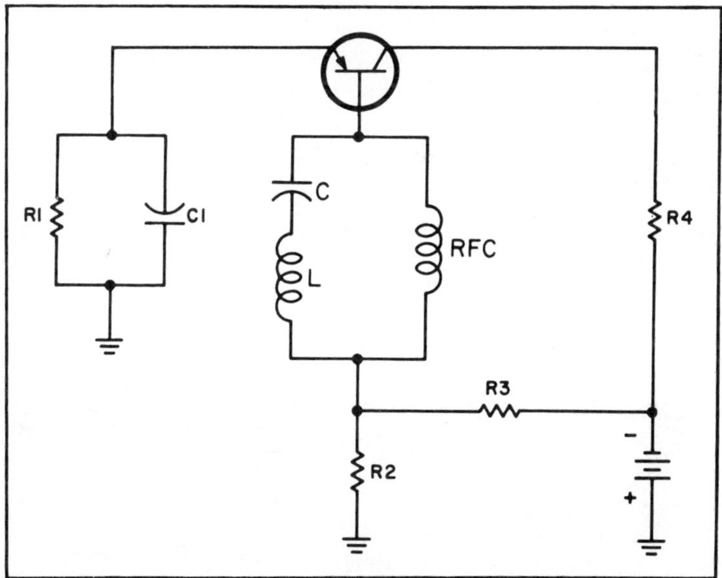

Fig. 10-16. The crystal-oscillator circuit becomes somewhat more understandable when the electrical equivalent of the crystal is substituted for it.

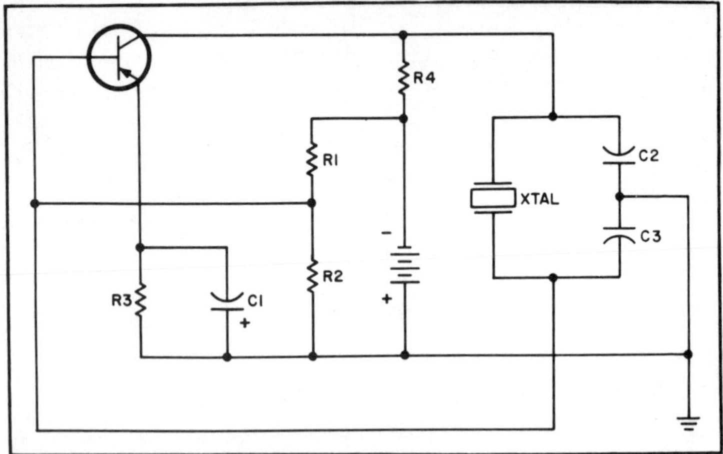

Fig. 10-17. This crystal oscillator circuit uses a capacitive voltage divider. Feedback is through C3 to the base-emitter circuit.

just serves to emphasize that without the choke we would have no dc return from the base of the transistor to ground—that is, to the plus terminal of the battery.

IF YOU CAN'T LICK 'EM—JOIN 'EM

The most positive cure for a headache is decapitation. This is quite a drastic cure for what may be a very minor illness. Similarly, the shunting capacitance existing between the metal plates of a crystal is usually nothing more than just a slight annoyance, particularly since the overall shunting effect of the circuit itself may be so much greater.

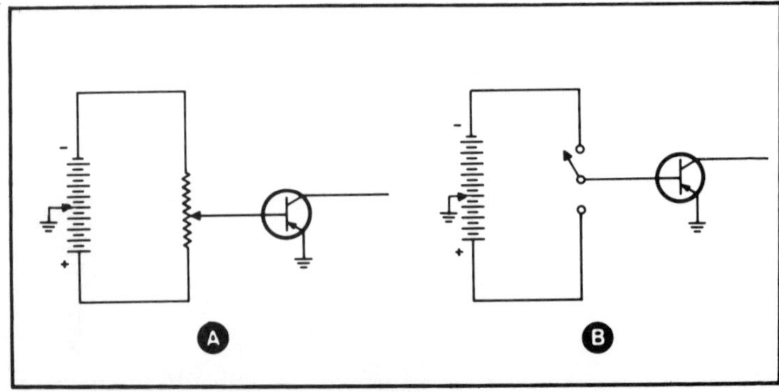

Fig. 10-18. In this arrangement, the transistor input is forward- or reverse-biased, depending on the position of the potentiometer arm or the switch.

Somewhat unexpectedly, we sometimes shunt capacitance directly across the crystal plates. Offhand, it would seem as though additional capacitance would be extremely undesirable, yet it can be made to work in a very beneficial way. We can use it to tune the crystal within small limits. Or, we can use this capacitance as an ac voltage divider. Or both. Figure 10-17 shows how.

A circuit analysis of this illustration shows how few new parts demand our attention. R3 and C1 are old-timers we've discussed in connection with other circuits. R1 and R2 are immediately recognizable as a voltage divider for forward biasing. R4 is the collector load. Other than the battery and the transistor, all that we have left in the way of components to study is the crystal and a pair of capacitors C2 and C3. C2 and C3 are in series, and this series combination is shunted directly across the plates of the crystal holder.

Now join us in a bit of mental transposition. In place of the crystal, visualize the electrical equivalent as shown earlier in Fig. 10-15. We can disregard the shunt capacitance since it will simply add to the equivalent capacitance of the series combination of C2 and C3. Similarly, we can forget about resistor R of Fig. 10-15. What is left? Just a series-connected inductor and capacitor in place of the crystal.

With the crystal in oscillation, an ac voltage at the resonant frequency will appear across the plates of the crystal holder. C2 and C3 are shunted across this voltage and will promptly divide it between them. Equally? Only if C2 and C3 are equal. If they are not of equal capacitance, the unit with the smaller capacitance will get the larger amount of voltage. Thus, we can control the division of voltage through proper selection of the relative capacitances of C2 and C3.

If you will trace the circuit you will see that the base is connected directly to the bottom end of C3 while the top end of C3 goes to the emitter. Offhand, this might seem incorrect, since, if you will move along from the top of C3 (on your way to the emitter), you will promptly run into C1. But the capacitance of C1 is so much greater than C3, that we can disregard it completely. Thus, if C1 is 1 μF and C3 is 100 $\mu\mu$F, the ratio is 10,000 to 1.

The voltage produced across C3 is fed back to the input since it is quite clearly connected in the base-emitter circuit. But the crystal and C2 and C3 are part of the collector circuit. In this way, a selected amount of output voltage, depending on the ratio of C3 to C2, is fed back to the input. And if we also arrange for proper phasing, we have met all the conditions necessary for sustained oscillation.

OSCILLATORS AND AMPLIFIERS

It is very nice to be able to put things into mental compartments. Designing a transistor circuit as an oscillator or as an amplifier falls into this category, but while it is convenient, it is also misleading. It tends to give us the impression that a wide chasm exists between oscillators and amplifiers. But is the separation as wide as we might think? I have an amplifier we can feed back part of the output, in phase, to the input. Would we then have an oscillator? Not necessarily. We would have a regenerative amplifier but it would not oscillate until the feedback reached a certain amount. Thus, the transition from amplifier to oscillator is not as clear and as sharply defined as we might have expected.

And now that we have managed to bring about an apparently closer relationship between oscillator and amplifier, let's separate them again. What is the difference? With an amplifier we select a particular operating point on the characteristic curve and that point determines our class of operation. An oscillator of the L-C type, though, runs the gamut from cutoff to saturation and back again. For an understanding of just what it is that a transistor does, then, we must always go to its characteristic curve.

TRANSISTOR AS A SWITCH

We can use selected portions of the transistor characteristic curve or all of it for amplifiers and oscillators, but if we wish we can also use the beginning of the characteristic (cutoff) and the end (saturation). Cutoff means no current while saturation is maximum current. This all-or-nothing-at-all state of affairs is what we get every time we use a switch.

Figure 10-18A shows our first step toward using the transistor in this function. When the arm of the potentiometer is at the top, the transistor is heavily forward-biased and we get maximum, or saturation, current. When the arm is at the opposite end, the base-emitter circuit is heavily reverse-biased and we are at cutoff. To get a faster transition from cutoff to saturation, we could substitute a single-pole double-throw switch for the potentiometer.

We want to use the transistor as a switch since it will be faster and much more dependable than manual switching. That is why the circuit of Fig. 10-18B, using a switch to control the bias, is self-

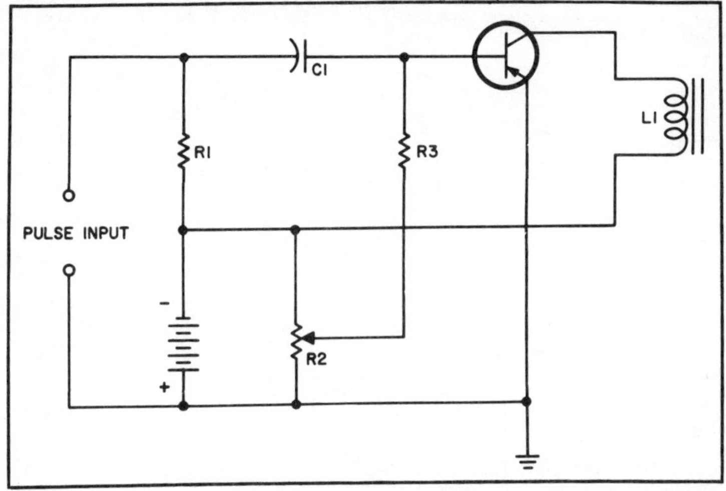

Fig. 10-19. Transistor switching circuit. The transistor has two operating points—saturation and cutoff.

defeating. But instead of a switch, we can control the biasing electronically, as shown in Fig. 10-19.

Let's analyze the circuit and see what we have. Since we have a pnp transistor we must have the emitter positive, the base, and the collector negative with respect to the emitter. Tracing the circuit shows that it meets these requirements, within limits. With the help of potentiometer R2 we can vary the amount of bias to the base-emitter circuit. When the arm is at the top, we get the maximum amount of forward biasing. Let us say that the voltage of the battery is sufficient to produce maximum transistor current—that is, saturation. This condition is comparable to having a closed switch.

A TINY ANTENNA

At the beginning of this chapter, a construction project was mentioned. Here it is! By now, you surely know enough about transistors to get something out of your knowledge. This antenna is really small. It demonstrates very well how the transistor can do remarkable things. You may also find this little antenna quite useful when it is inconvenient to put up a full-size short-wave antenna for receiving signals from 3 to 25 MHz.

You don't need to be led by the hand in the construction process, so we won't discuss things like the size of the cabinet you should use (although it can be pretty small!), or when to drill the

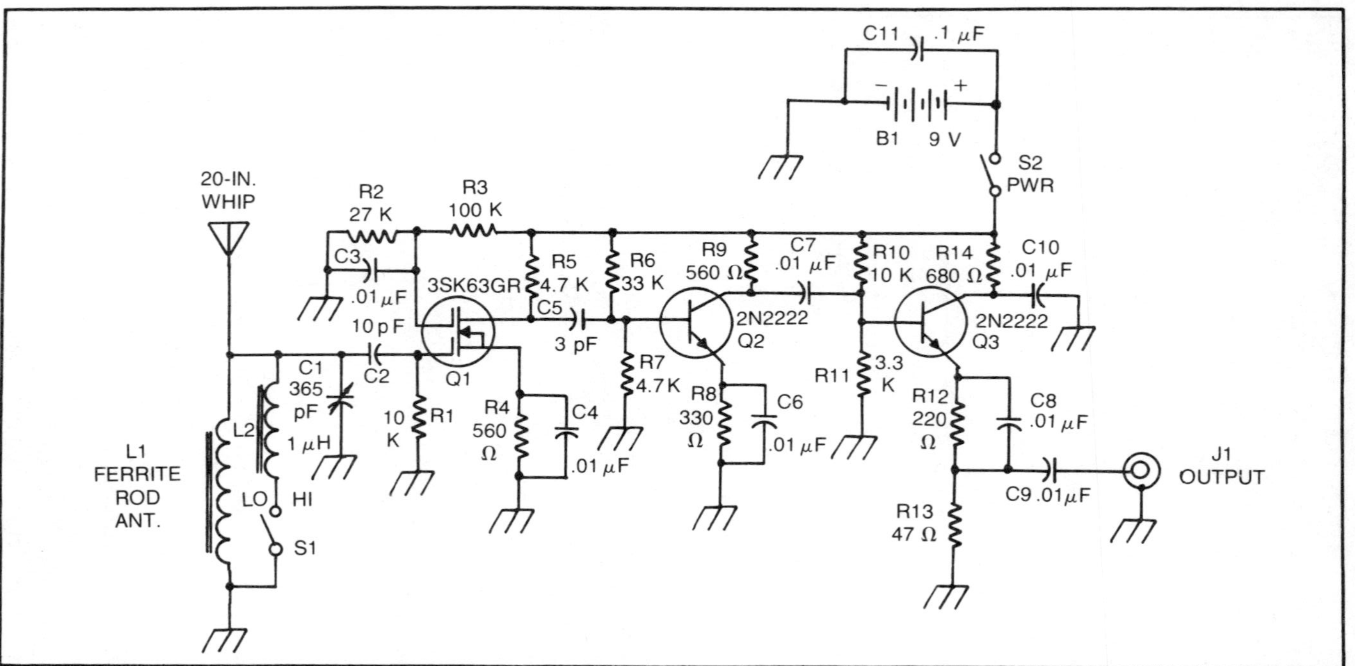

Fig. 10-20. Circuit diagram of the miniature shortwave antenna. Resistances are in ohms; K = 1000. Capacitances are as shown. All are disk ceramic except for the variable, C1.

holes for the switches and things. The cabinet should be made of plastic and not metal, if you plan to put the ferrite-rod antenna inside the enclosure! But other than this constraint, there are none.

The complete circuit is shown in Fig. 10-20. It looks a little more complicated than the other circuits in this book. Actually, though, as with any circuit, this device can be broken down into stages. In this case, there are three: two amplifiers (Q1 and Q2) and an impedance-matching stage (Q3). Let's go through the circuit stage by stage, as a short-wave signal does.

THE FERRITE ROD

The ferrite-rod antenna is a story all by itself. This antenna is smaller than your little finger. How can that be expected to perform as well as a longwire? Well, it has to do with the effect of ferrite material on magnetic flux. Because of the stick of ferrite, the coil acts as a much larger loop than its dimensions would lead us to believe. We won't get into a lengthy discussion of wavefronts and magnetic lines of force. But it is quite amazing how this little rod works. The rod used in this project has a permeability of 125, and consists of an Amidon type R61-050-400 ferrite stick broken in half. To break the rod, nick it in the center with a triangular file or similar instrument to a depth of about one millimeter, and break it with your fingers. Wind 15 turns of enameled wire on this rod. The wire size is not too important. If a core of different permeability or physical size is used, the number of turns needed will be different. Experimentation will be necessary to get the desired tuning range.

On the higher frequencies, an inductor of about 1 μH is switched in parallel with the ferrite loopstick. Also, a short whip antenna, connected to the ungrounded side of the loopstick, enhances the pickup above 10 MHz.

THE FRONT END

Once the signal has entered the antenna system, it is either shunted to ground (if the circuit is off resonance) or passed on to Q1 (if the circuit is at resonance). Tuning is accomplished by a variable capacitor, C1, and the range switch S1. At resonance, the signal goes to the first gate of the MOSFET. Here, the signal is amplified about 25 dB. The input impedance of the circuit is extremely high. The output impedance is also quite high, but smaller than the input impedance. The dual-gate MOSFET provides plenty of gain, even well up past 25 MHz. But the antenna system is designed to go only to 25 MHz. Above that point it loses efficiency.

It is essential that the front end of any receiver be linear, and this preamplifier is no exception. An extremely strong signal off the resonant frequency of the antenna should not cause the front end to go crazy. The tuned circuit has an extremely high amount of selectivity. Without it, local broadcast signals (among others) would drive the circuit into nonlinearity, and the result would be disaster!

THE SECOND AMPLIFIER AND EMITTER FOLLOWER

From the dual-gate MOSFET, our signal passes to the broadband amplifier Q2. This stage has about 20 dB gain at 3 MHz, dropping to around 10 or 15 dB at 25 MHz. Together with the dual-gate MOSFET, then, we can see that there is a lot of amplification!

The coupling capacitor between the MOSFET and the npn bipolar transistor Q2 is small, and for good reason. Because of the greater gain of this circuit (and also the greater amount of selectivity) at the low frequencies, there must be some rolloff at the low end of the short-wave spectrum. A series capacitor does this job nicely. The rolloff begins at about 10 MHz and gets greater as the frequency goes down. Without this capacitor, the circuit might break into oscillation at the lower frequencies. The gain is still more than enough at 3 MHz with the rolloff.

The output of the second amplifier has an impedance in the neighborhood of 500 ohms. Most communications receivers, however, want to see 50 ohms at their antenna terminals. The third stage of our circuit, Q3, is an emitter follower (common collector). It has no gain; in fact, there is a little bit of loss. But the improvement caused by proper impedance matching is far more significant than the slight loss in the transistor. The emitter resistors determine the output impedance.

GETTING THE PARTS

The ferrite rod can be obtained from Amidon Associates, 12033 Otsego Street, North Hollywood, California 91607, although this is certainly not the only source of ferrite rods. The dual-gate MOSFET may be a type 3SK63GR, GE FET-4, Sylvania ECG-221, or equivalent. The rest of the parts—resistors, capacitors, npn transistors, and 1 μH inductor—can be found in most Radio Shack stores. The 1 μH coil deserves a little extra attention, since it is a modified part. Radio Shack supplies 10 μH coils on solenoidal forms with axial leads. You can easily change the inductance to 1 μH by cutting the coil wire from one of the axial leads and removing 2/3 of

Table 10-1. Parts List for the Miniature Shortwave Antenna.

Description	Reference	Other Info
Battery, 9-volt	B1	
Battery holder	-	
Cabinet, plastic	-	
Capacitor, 3 pF	C5	Disk ceramic
Capacitor, 10 pF	C2	Disk ceramic
Capacitor, 365 pF	C1	Calectro AL-232
Capacitor, .01 μF	C3, C4, C6, C7-C11	Disk ceramic
Capacitor, .1 μF	C11	Disk ceramic
Circuit board	-	Radio Shack 276-160
Cord, phono	-	
Ferrite rod	L1	Amidon R61-050-400 (Break in half)
Inductor, 1 μH	L2	Modified Radio Shack 10 μH
Jack, phono	J1	
MOSFET, dual-gate	Q1	3SK63GR, GE FET-4, Sylvania ECG-221 or equivalent
Resistor, 47 ohm	R13	Carbon 1/4-watt
Resistor, 220 ohm	R12	Carbon 1/4-watt
Resistor, 330 ohm	R8	Carbon 1/4-watt
Resistor, 560 ohm	R4, R9	Carbon 1/4-watt
Resistor, 680 ohm	R14	Carbon 1/4-watt
Resistor, 3.3 K	R11	Carbon 1/4-watt
Resistor, 4.7 K	R5, R7	Carbon 1/4-watt
Resistor, 10 K	R1, R10	Carbon 1/4-watt
Resistor, 27 K	R2	Carbon 1/4-watt
Resistor, 33 K	R6	Carbon 1/4-watt
Resistor, 100 K	R3	Carbon 1/4-watt
Switch, SPST	S1, S2	Miniature toggle
Transistor, 2N2222	Q2, Q3	npn bipolar
Whip antenna	-	Stiff wire, 15 to 20 inches
Wire, enameled	-	No. 22, 2 feet
Wire, hookup	-	No. 22 insulated stranded, 2 feet

the coil turns. Then, cut the coil wire, scrape off the enamel, and resolder the cut end to the axial lead. The new coil will have 1/3 the turns, and hence about 1/9 the inductance, of the old coil.

A complete parts list is given in Table 10-1. Included in this list is a circuit board ideal for experimenter use. It has a foil backing that groups the holes in fours. This board has 20 rows and 20 columns of holes, labeled A through T and 1 through 20. The board can be cut between rows 11 and 12. A suggested component layout for the circuit is shown in Fig. 10-21. (Of course, you may use any kind of circuit board, even nails pounded into a block of wood, if you wish!)

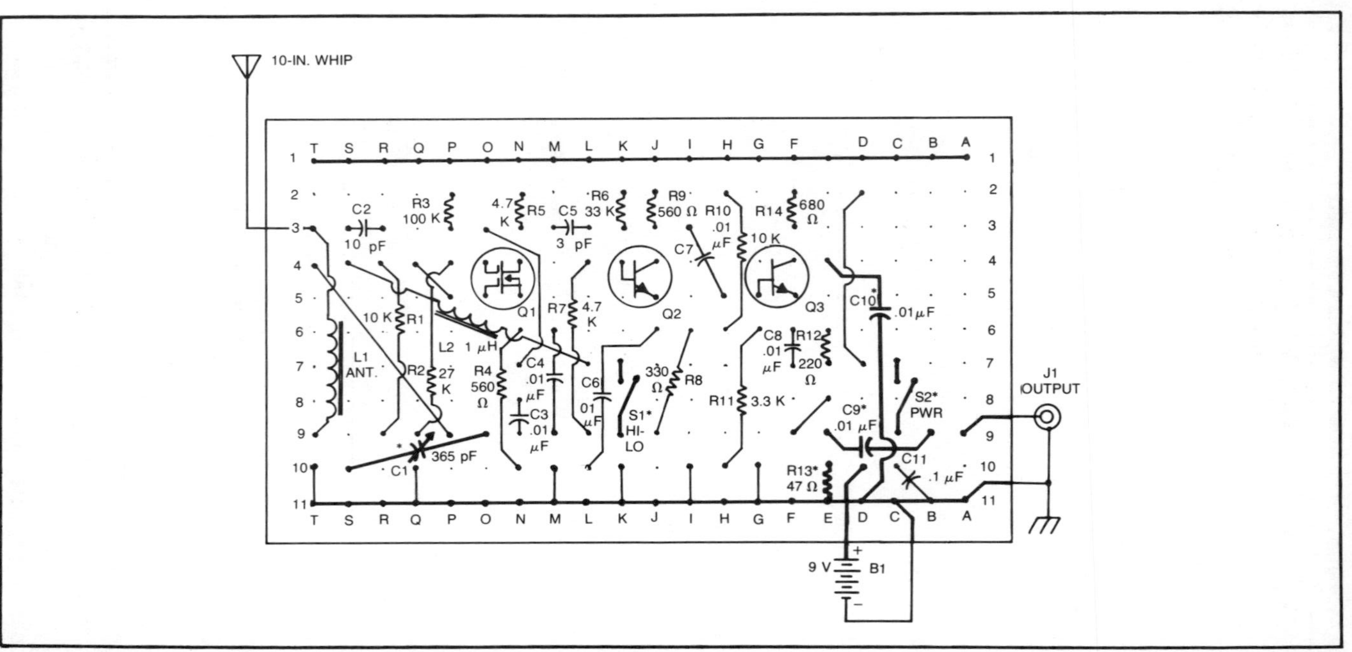

Fig. 10-21. Suggested parts layout for the miniature shortwave antenna using Radio Shack experimenter board. This view is from the nonfoil side of the board. Components in light black are mounted on the nonfoil side; components with an asterisk (*) and in heavy black are mounted on the foil side. Plain lines indicate jumpers. A connection is indicated by a dot.

ASSEMBLY

Assuming you use for the Radio Shack board, connect the jumper wires first. Use insulated hookup wire, about No. 22 or 24 gauge. Put the 15-turn ferrite loop in place firmly against the nonfoil side of the circuit board. Attach a stiff, 15- to 20-inch wire whip to the ungrounded end of this loopstick.

Next, install the resistors. When a resistor is connected between two adjacent holes, put the resistor in vertically, bending the top lead around 180 degrees.

Install the capacitors after the resistors. Connect the 365 pF variable to the circuit board using hookup wire. Make the leads as short as possible. Calectro makes an excellent receiving-type variable for this purpose, which can be soldered right to the foil on the circuit board. Solder it in the best position for front-panel mounting. The SPST switches should be soldered right to the circuit-board foil as well, if possible. Tin the lugs and the foil and heat them together. If the 365 pF capacitor is not connected directly to the board, you might want to put the switches away from the board. Use hookup wire, in lengths as short as possible. Connect the 9-volt battery holder. Be sure the polarity is right! Connect the output jack; be sure the inner conductor doesn't go to ground!

Finally, put in the transistors. Do not use too much heat. Even silicon can be destroyed if you try hard enough. Be especially careful when handling the dual-gate MOSFET, as it is easily damaged by static discharges. It's not a bad idea (as silly as it sounds) to tape a well-grounded piece of wire to both wrists when you handle one of these fragile things.

Check your wiring carefully. It might save you some grief later. Are the transistors connected right? Do you know which lead is which? Are all the other components going to the right holes? Are there any solder bridges, cold or unsoldered connections, or broken foil points on the foil side of the board? Are all components accounted for? If so, you are ready to hook this little antenna to your receiver and test it.

TESTING

Use a shielded phono cable to connect the tabletop antenna to your short-wave receiver. Tune in a frequency of about 5 MHz if it is nighttime, and 15 MHz if it is daytime. The high-frequency range is about 10 to 25 MHz, so if you tune your receiver to 15 MHz, set C1 to about the middle of its range, and set the HI/LO switch, S1, to HI. The low-frequency range of the circuit is approximately 3 to 10

MHz; of you want to hear signals at 5 MHz, set C1 at the middle of its range and S1 to LO.

Adjust C1 until a rise in the noise level is heard. This will probably be more apparent at 5 MHz than at 15 MHz, since the lower frequencies are noisier by their very nature. The peak should occur fairly close to the middle of the range of C1. Tune the receiver to a signal and peak the level carefully with C1. You should observe a sharp effect.

Now, let's suppose for a moment that the worst happens: nothing! What should you do? It certainly can be frustrating after you have spent several hours on a project, and find that it works as well as a pack of cigarettes. First, be sure the receiver is working! Try it with your old antenna, or a piece of wire about 50 feet long. If your receiver is all right, then you know (of course) that the problem is with the tabletop antenna.

The most frequent causes of an inoperative new circuit are as follows:

1. No power to unit.
2. Wiring errors or omissions.
3. Defective component or components.

You can pretty much figure out how to solve each of these. We cannot get into a complete technician course in this book. But this circuit is a simple one. There aren't too many things to go wrong with it. The worst possible thing that can happen is a component value error, such as a resistor that's off by a decimal point. (This sometimes does happen!) A shorted or open capacitor can also be pretty difficult to locate.

It's up to you to put the finishing touches into this project. If you use a cabinet enclosure, be sure it's plastic and not metal. A cabinet with an aluminum front panel and plastic sides, top, and bottom and back is all right. A metal enclosure will shield the ferrite-rod antenna and render it utterly worthless. Don't put this circuit inside a receiver with a metal cabinet, either! Other than these precautions, which are essential, there is nothing critical about finding a good place for your new antenna.

FINALE

We have come to the end of our story. But yours need not finish here. This book should be like an appetizer (hopefully) to a ten-course dinner for you—a continuing interest in solid-state technology. If you have followed this text, you are ready to move on to a more detailed study of transistors. New kinds of transistors are constantly being developed, and some professionals spend a great deal of time just keeping their knowledge of them up-to-date! You certainly don't face a shortage of interesting experiments or projects with transistors.

The scientists at Bell Telephone Laboratories must have known they had "a good thing going" in 1948. But even they probably didn't realize, at the time, the magnitude of the impact their transistors would have in electronics. Transistors, in some form, are in almost all the electronic devices we now take for granted.

QUESTIONS

1. Describe the operation of a meter amplifier. What is an advantage of using such a circuit?

2. Draw a schematic diagram of a simple transistor-operated relay amplifier. How does it work?

3. What precautions must be taken when connecting electrolytic capacitors in a circuit?

4. What is a signal tracer?

5. If a circuit diagram is drawn so that you cannot immediately understand the functioning of some of the components, what can you do to overcome this difficulty?

6. Explain how a transistor can operate as a voltage divider.

7. Draw a common-collector amplifier circuit and explain how it works.

8. Draw a multivibrator circuit and explain its operation.

9. What is a transducer? Give three examples.

10. Draw a circuit diagram of a crystal-controlled transistor oscillator.

11. What is the purpose of a shunting capacitor across a battery?

12. What are typical values of coupling capacitance used in audio transistor amplifiers?

13. Are the transformers used in transistor audio amplifiers stepup or stepdown types? Explain.

14. What is the advantage of doing a circuit analysis?

Index

Index

D

E

F

G

H

I

J

L

V

Z

Edited by Roland S. Phelps